MAKING YOUR OWN MOTOR FUEL

With Home and Farm Alcohol Stills

FRED STETSON

GARDEN WAY PUBLISHING
Charlotte, Vermont 05445

To Marc Rogers

Illustrations by Cathy Baker

Printed in the United States

Library of Congress Cataloging in Publication Data

Stetson, Fred, 1943–
   Making your own motor fuel, with home and farm alcohol stills.

   Bibliography: p.
   Includes index.
   1.   Alcohol as fuel.   I.   Title.
TP358.S73   662'.669   80–19640
ISBN   0-88266-163-9 (pbk.)

Contents

Introduction

Each week, interest in alcohol fuel accelerates. Alcohol fuel associations are forming. Colleges and companies offer workshops on fuel making. For those who want to build large plants, the federal government may offer grants worth thousands. New manufacturers, suppliers, books, pamphlets or plans come on the market every month. The stock of the nation's leading alcohol producer, Archer Daniels Midland Co. of Decatur, Illinois, more than tripled in one year. And the National Alcohol Fuels Information Center says it receives 500 telephone inquiries a day.

Where is this all going and where will it end? It's hard to say. But every time the price of gasoline increases at the pump, the future for alcohol fuel seems more and more promising. Someone suggested awhile ago that the alcohol fuel industry may be at the threshold of a massive expansion. Are we at this point? Are we beginning to relive the days when Henry Ford put thousands of motorists on the road in Model Ts? In a few years will backyard stills be commonplace?

It's too early to tell. But, if you want to share this interest, if you want to have a "taste" of alcohol fuel making, this book will help you get started. With it, you can learn how to build and operate two small stills. And you can learn about the operation of two larger, farm-sized plants, one owned by the Zeithamers of Minnesota and the other by the South West Alabama Farmers Cooperative Association. For a bit of perspective from the backwoods, there's Joseph Earl Dabney's colorful chapter on moonshining. He calls it "The ABCs of Pure Corn." Other chapters explain automobile engine modifications and federal regulations, which are simple for the backyard experimenter.

Since work was first begun for this book several months ago, the technology of small-scale fuel making has developed at a rapid pace. One knowledgeable individual told me the advancements of the past 18 months have exceeded the advancements of the previous 10 years. So, if you want to keep up with this, and you are serious about making large quantities of alcohol fuel (about 50 gallons a week or more), let me suggest the following:

Read as many books on the subject as possible; study and compare different plans and designs; visit an alcohol fuel plant with the performance you hope to duplicate; and ensure that any still you contemplate building or purchasing performs the way its designers or sellers promise. Those who want to produce *large* quantities of fuel should watch the marketplace. Look out for advancements in distillation technology. There may be a reasonably priced still that's right for you.

But, for the pioneer, for the person who wants some "hands-on" experience at little cost, this book is a guide for starting and, I hope you agree, a valuable source of information. I also hope my desire to provide information in a straightforward manner does not fully mask the sense of enjoyment I've had, assembling a small still and operating it on my kitchen stove, learning the basics of alcohol fuel making, visiting with a "revenuer" from the Bureau of Alcohol, Tobacco and Firearms, and, last but not least, marveling at the sight of blue flame as I touched a match to my own home-made motor fuel.

In sum, this is a guidebook for the small backyard experimenter and an introduction for the person contemplating larger operations. All the information needed to build and operate two small stills is here. Just as important, there are complete lists of suppliers and their addresses. For the small-scale distiller, that's more important than it might seem at first. In many parts of the country, distilling supplies are not readily available. But you can obtain them by simply writing for catalogs and placing an order. For those with larger operations in mind, there are names and addresses of many manufacturers, suppliers, engineers and sources of help. If you have any questions, write or call the:

National Alcohol Fuels Information Center
1617 Cole Boulevard
Golden, Colorado 80401
1-800-525-5555 (toll free)

CHAPTER 1

About Alcohol Fuel Making

Making alcohol's simple. It's as old as the hills of Georgia, Alabama and Kentucky — states where "makin' 'shine" is embedded in folk tradition, if not in the pages of history books. Why, all you do is take a little corn, boil it, throw in some yeast, let it set for a few days, and then cook it. Alcohol vapors rise and wind through a coiled tube and, before long, "likker" is spewing out the end of the coil. Run it through a couple of times and you get a high-proof alcohol, good enough for a stiff (but sometimes poisonous) drink and, with a few changes here and there, good enough to run a car or tractor.

Well, unfortunately, it's not quite that simple.

It is true that you can make a few gallons of alcohol *fuel* right in your own kitchen, using a few simple, readily available implements. And it is true that this is both legal and fun. But small-scale alcohol fuel making from corn or other suitable vegetables can also be time-consuming and expensive. Take, for example, the cost of corn, a popular raw product. Most experts say you can get up to two and a half gallons of alcohol from every bushel of corn. But in some parts of the country, corn — shelled and ground for alcohol making — costs $3.65 a bushel. In other words, for some, the raw product alone costs $1.46 a gallon. To reduce this cost, perhaps you can grow your own corn or another suitable vegetable.

Advantages of Making Fuel

While considering costs carefully, remember that there are advantages to making alcohol fuel at home. The economics of it may be in your favor. This is especially likely if you raise your own vegetables for alcohol fuel making. The costs of gasoline and other conventional fuels are unlikely to decline. Furthermore, home production of alcohol would give you an alternate fuel supply. By-products from alcohol making with corn may be

1

used as a high-protein livestock feed. You may be able to use waste corn stalks, wood scraps and other inexpensive materials as a fuel to power your alcohol still.

Making Alcohol Fuel

Ethyl alcohol, also known as *ethanol*, is the kind of alcohol I'm going to be talking about. It may be made from many vegetables and other agricultural produce. Corn and grain are a few of the suitable *starchy* materials. Others include oats, rye, wheat, sorghum, potatoes and Jerusalem artichokes. The alcohol fuel recipes in this book call for corn. Suitable *saccharine* materials, those with a sugar content, include sugar cane, sugar beets, molasses and fruit juices. In the United States, molasses and grains are the principal materials used.

Materials such as sugar cane are first pressed and then the sugar-rich liquid is fermented, with the help of yeast, to produce alcohol and carbon dioxide. If starchy materials such as corn and other grains are used, enzymes or other agents must be added to help break down the starch and convert it to a fermentable sugar. In either case, an aqueous solution with an alcohol content of at least 10 percent is the desired end result of fermentation.

Making alcohol is a strong tradition in the south, but when the product is an illicit beverage, rather than a fuel, the "law" intervenes. Here, revenuers have strapped a confiscated still to their car.

After fermentation, this solution, prepared in vats of various sizes (as small as a kitchen pot, as large as several thousand gallons), is then carefully heated to a temperature warm enough to cause the alcohol to vaporize, but low enough so the water boiling point won't be reached. This process of separating the alcohol from the liquid solution is called *distillation*. (The word *still* is an abbreviated version of the longer word.) In the final step, the alcohol vapors condense, typically inside water-cooled tubing, and then return to a container or collector.

Alcohol and Gasohol

What you've just read is a simplified description of making alcohol at home or on the farm. The end result, after one or two distillation runs, should be an alcohol with at least 160 proof (80-percent alcohol, 20 percent water). This is adequate for combustion engines, but unsuitable for blending with gasoline. *Gasohol* is a blend of unleaded gasoline and high-proof, water-free alcohol. This alcohol cannot be produced with most home-built, small-scale stills. But the alcohol you *can* produce with your own still is strong enough to be used in engines, with minor modifications.

Alcohol with 160 proof is 80 percent alcohol and 20 percent water (left). Gasohol is 90 percent unleaded gasoline and 10 percent alcohol (right).

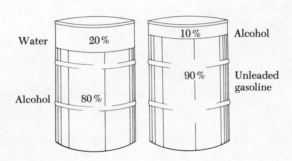

Other Items to Consider

Although fermentation and distillation techniques are well known, it takes some practice to do them well. It's a good idea to think small, to start with a small still — and then to have the patience of a good cook. "It's like a woman making biscuits," "Short" Stanton, an ex-distiller from Tennessee told Joseph Earl Dabney, the author of *Mountain Spirits.** "If she

Mountain Spirits: A Chronicle of Corn Whiskey from King James' Ulster Plantation to America's Appalachians and the Moonshine Life by Joseph Earl Dabney (Lakemont, Georgia: Copple House Books, 1978).

don't know how to mix that dough in the bread over thar, when she puts 'em in the pan, they ain't no count."

The ways that fermentation and distillation are achieved are many and varied. There are several books, plans, manufacturers and companies that suggest different techniques and equipment. There are multimillion-dollar stills; there are portable stills costing $12,000; and there are stills built from junk materials that cost almost nothing. In this book, two simple alcohol fuel-making procedures and stills are described in detail. Two larger farm operations are also described. And the appendices include the names of engineers, manufacturers, suppliers and dozens of other sources of help and information.

Large Stills

You can assemble and operate a small, five-gallon still described in this book with little difficulty. Building and operating the 55-gallon still described here takes more time and effort. If you are thinking of producing, say, 25 gallons a week or more, review Chapters 4 and 5 on the Zeithamer and Turner stills. Then, if you're still interested in producing large quantities of alcohol fuel, I'd suggest studying many different books, plans and working stills. There are two books that I recommend for farmers or others who plan to produce large quantities of fuel. These are:

A Learning Guide for Alcohol Fuel Production, produced by Colby Community College in Kansas.*

Fuel From Farms, A Guide to Small-Scale Ethanol Production, by the Solar Research Institute, Golden, Colorado.*

Examine these and other books before making a final decision on the equipment and operation best suited for you. If you're thinking of large-scale production, one way to start would be simply to ask yourself how much fuel you anticipate making in a year, month, or week, and then design the still best suited to meet your needs. Even if your ultimate aim is to build a large-capacity plant, start with smaller equipment, especially if you haven't had previous fermentation and distillation experience or if you can't get professional help.

Another important consideration for the large producer is the availability and suitability of raw materials. What agricultural produce is readily available that can be transported and processed cheaply? Determine

---

*For complete names and addresses of publishers, see Suggested Additional Reading.

**FIGURE 1-1. POTENTIAL ANNUAL ALCOHOL YIELDS
PER ACRE OF BIOMASS RESOURCES***

Biomass resource	Unit of yield	Average yield per acre	Gallons alcohol per unit of yield Ethanol	Average gallons alcohol per acre Ethanol
GRAINS:[1]				
Corn	Bushels	91	2.6	237
Wheat	Bushels	31	2.7	84
Grain sorghum	Bushels	56	2.6	146
SUGARS:[1]				
Sugar cane	Tons fermentable sugar	4.6	136	626
Sugar beet	Tons fermentable sugar	2.6	136	354
Sweet sorghum	Tons fermentable sugar	2.8	136	381
SUGARS:[2]				
Sugar cane	Tons fermentable sugar	7.6	136	1,034
Sweet sorghum	Tons fermentable sugar	4.4	136	598
AGRICULTURAL RESIDUES:[1]				
Corn stover	Dry ton equivalent	2.0	47	94
Wheat straw	Dry ton equivalent	2.4	47	133
WOOD:				
Silviculture farm	Dry ton equivalent	8.0	47	376

*Adapted from "The Report of the Alcohol Fuels Policy Review," U. S. Department of
Energy, June 1979.
[1]Based on current average yields per acre.
[2]Based on projected yields for the year 1995 assuming narrow row spacing, fertilization, and
  geographically tailored varieties.

the energy costs associated with the production of the *crops* as well. If
amortization of the alcohol-making equipment depends upon the use of
an inexpensive product, what is the long-term price outlook for that product? For an idea of the *relative* yields of various raw products, see Figure
1-1.

Examine the energy costs of processing and cooking the product. What
fuel will you use to generate the heat needed for fermentation and distillation? These are questions to ask and consider, even if your alcohol-making
operation is a small one. For larger operations, it would do no harm to
prepare a budget, outlining all the costs of building and operating your
plant. Budget your time, too.

Another important step is to consider the ways to use your alcohol fuel
and its by-products. If the alcohol is to be run "straight" in your motor vehicles or engine-driven equipment, what will be the effect on the engines?

The most popular and common ingredient for alcohol fuel making is corn.

This North Carolina still had coke-fired boilers and, at right, a "thumper keg" for second-stage distilling. Revenuers stand watch.

Will warranties remain in force? What are the manufacturers' recommendations? Would these by-products, such as high-protein corn residues, be easy to add to your livestock feed? Or would it be necessary to dispose of them or transport them away from your home or farm?

Federal Regulations

Alcohol fuel can be made safely and legally at home or on the farm. The Bureau of Alcohol, Tobacco and Firearms, with regional offices across the country, has jurisdiction over the operation of stills. Obtaining permission to operate a small, experimental backyard still is *not* complicated. (For larger, commercial plants, the paperwork *is* more extensive and complex.) Basically, the BATF asks the backyard experimenter to write a single, detailed letter, explaining what he or she plans to do. But you should also familiarize yourself with state and local environmental, safety, and zoning regulations. More detail on the federal BATF requirements is given in Chapter 8.

In these pages, many questions and considerations have been raised. It has been the intent not to discourage you from making alcohol fuel, but to raise some key points worth considering at the outset. In the following chapters, these and other questions will be answered, and enough information will be presented to help you plan, design and build a small, workable still.

CHAPTER 2

Building and
Using a Five-Gallon Still

After you decide to make alcohol fuel, consider the scope of your operation and the size of your still, and then select a suitable site. Regardless of the still size, use an easily accessible site with running water, a source of heat and a place to dispose of wastes. For a small, five-gallon still, a kitchen is perfectly adequate, as long as your spouse shares your enthusiasm for "home-brewed" fuel. Keep in mind, too, that mash ferments at about 70° to 90°F., so try to use a place where mash temperatures within this range can be maintained.

If you're not familiar with mash preparation, fermentation and distillation, it's best to gain experience with a small still. No matter how much you may have read about these processes, many people agree that you can't really learn them without doing them. It helps to know what fermented mash looks like. So, to avoid costly mistakes, start small. Try a simple still fashioned from readily available parts and materials. Shown is a small still with these basic parts:

- A 5-gallon pressure cooker, used for mash cooking and distillation.

- A 15-foot-long piece of ⅛-inch copper tubing, coiled and placed in a cooling tub where it acts as a condenser.

- A 6½-gallon cooling tub or "flake stand," in this case a tin-plated lard can.

- A small container to collect the alcohol.

- Two garden hoses to provide cold water to the flake stand and to return the coolant to a sink or drainage point.

9

A small, five-gallon still may be fabricated from readily available materials and then set up for distilling on nothing more than a kitchen stove.

Other durable containers with similar capacities may be substituted for the pressure cooker and lard can. If you fabricate a cooking container, make a removable, cone-shaped lid that "channels" the alcohol vapors to the copper tubing.

The basic parts of the five-gallon still are garden hoses, condensing coil, condenser or flake stand and pressure cooker.

How It Works

Briefly, mash cooking and distillation in this small still works this way:

- The mash, a mixture of cornmeal and water treated with enzymes, is heated to specific temperatures, then cooled.

- The mash ferments for three to seven days.

- The mash is reheated in the pressure cooker until it reaches temperatures warm enough to produce alcohol vapors.

- The vapors rise from the cooker, pass through the copper tubing and into the coil, where they condense and flow into a collector as alcohol distillate.

More information on these steps follows, but first let's build a still.

PUTTING TOGETHER A SMALL STILL

Remember, there are few hard, fast rules about homemade distillation apparatus. A small still may be assembled easily from parts and materials mentioned here. But you may have equally suitable substitute materials with which you can improvise an equally suitable still. Furthermore, it's not critical that dimensions discussed in this chapter be duplicated precisely.

Cooker

A five-gallon pressure cooker, with a few adaptations, is an ideal container for cooking and distilling mash. It also may be used for fermentation, but it's cleaner and simpler to use a separate container for fermentation; that way, your cooker is not tied up while fermentation progresses.

A good cooker is the *All-American*, manufactured by the Wisconsin Aluminum Foundry.* This cast aluminum model has a dial-face pressure-temperature gauge and a small metal pivot that acts as both pressure control and safety valve. Use the cooker *as is* during cooking. During distilla-

---

*See Appendix A for a complete list of materials, suppliers and manufacturers.

The metal pivot to the left of the lid handle is removed and replaced by a copper fitting that attaches the cooker to the copper condenser tubing.

tion, remove the pivot and then screw appropriately sized fittings into the remaining threaded hole. The fittings connect the pressure cooker to the copper tubing.

Copper Tubing

Coil the soft copper tubing (⅛ or ¼ inch in diameter) tightly to achieve the best condensation of alcohol vapors. The deeper and wider the cooling tub, the greater length of copper needed to provide maximum cold surface exposure. About 15 feet is adequate for a tub with about five gallons capacity.

To coil the tubing, find a large, round object, slightly smaller than the inside diameter of the flake stand. Slowly and carefully wrap the tubing around the object in neat coils. A friend, Chuck Armel, made my coil by wrapping the tube around a one-gallon paint can. It works fine.

It's easier to coil the copper tubing if it is first heated with a butane torch. Also, if the tubing has a diameter greater than half an inch, some moonshiners suggest plugging one end and pouring dry sand in the other to prevent crimping during wrapping. (I have not found it necessary to do this — and I doubt that I'd try it. Seems to me there's a good chance sand could foul and possibly block the tube.)

Remember to leave about 18 inches of the tubing straight. This length will extend toward the pressure cooker and be attached with the small fittings.

MATERIALS FOR FIVE-GALLON STILL

1 five-gallon pressure cooker or similar capacity container, preferably with a lid having two outlets

1 piece of soft copper tubing, ⅛ or ¼ inch in diameter and about 15 feet long

1 copper compression union

1 threaded fitting

2 garden hoses long enough to reach from still to water source and drainage point

1 male garden-hose fitting

1 female garden-hose fitting

1 reel of solder and a butane torch

1 cooling container or flake stand with a capacity of about five gallons

1 small plastic container to collect alcohol

1 adapter to convert a kitchen sink faucet to accommodate garden hoses

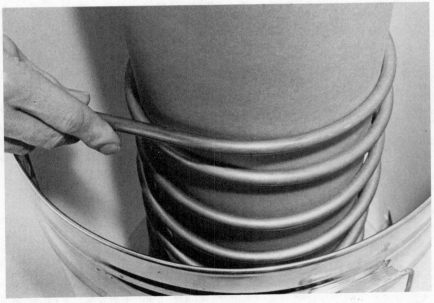

Copper tubing is wrapped around a round object to form a condenser coil. The coil will be secured inside the cooling tub.

Flake Stand

Flake stand is a term often used by moonshiners to refer to a tub or large container of water used to cool the copper coil and thereby condense the alcohol vapors into a nearly clear distillate. To transform a lard can or some other container into a usable flake stand, start by drilling or cutting three small holes:

- Cut one hole near the bottom of the can where the end or "tail" of the copper coil will exit.

- Cut a second hole near the bottom, large enough for a garden-hose fitting. Water will enter through this point.

- Cut a third hole near the top of the can, about five inches below the lip. This hole should also be the size of a garden-hose fitting; water will exit here.

After cutting these holes, solder the garden-hose fittings in place. Use a *female* fitting near the base of the flake stand to accommodate the *male* end of a garden hose. The female end of the hose may be screwed onto a sink faucet with an adapter, or any other appropriate faucet.

Solder a *male* garden hose fitting below the lip of the flake stand. It is

A flake stand with copper coil exiting from lower front. Garden-hose fittings are soldered at holes cut into the lower left and upper right sides of this cooling tub.

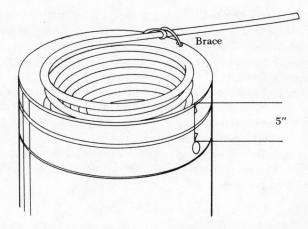

Cut a hole about five inches from the lip of the flake stand. Solder a female garden-hose fitting here. Note small brace that holds coil in place.

located about five inches below the lip to ensure that water is drawn off before the tub fills to the brim. I soldered my fitting too near the top, and water overflows unless I watch the tub closely.

Then, solder the copper coil so the straight 18-inch section exits from the top and the "tail" exits through the bottom. To brace the coil, wrap a small piece of copper wire around the straight section and then solder the wire to the lip of the flake stand.

One final word of caution: Once the coil is soldered in place, inside the flake stand but not yet attached to the pressure cooker, blow through it to make sure that it is free of dirt or any other obstruction. Be sure you feel air pressure coming out of the coil's end. Be careful that you don't twist the copper or grab it abruptly—doing so might cause the coil to dislodge. Also, as a last precaution, fill the flake stand with water and check for leaks.

WRITING THE BATF

At this stage, sit down and write a letter to the federal Bureau of Alcohol, Tobacco and Firearms, explaining what you're doing and applying for an experimental distilled spirits permit. See Chapter 8 for the toll-free telephone number and address of the BATF regional office nearest you and for additional information about what needs to be included in your letter. *It is illegal to distill alcohol, even if used as a fuel, without a federal permit.* If you are undecided about how much alcohol you intend to produce, read through the entire book once, at least quickly, to learn about various stills. When I wrote to the BATF, I applied for permission to operate both 5- and 55-gallon stills. The BATF approved this request. The bureau imposes the same requirements on all experimental still operators who produce less than 5,000 gallons annually.

Keep in mind that because there has been a dramatic increase in the number of applications for BATF experimental distilled spirits permits, it takes time to receive one. Furthermore, the bureau is taking steps to simplify and streamline the process; this means the rules may change by the time you read this book. I'd recommend a phone call to the nearest BATF office to be sure that you have the latest information on the bureau's requirements and policies.

USING A FIVE-GALLON STILL

Corn Mashing

There are dozens, perhaps hundreds, of ways to prepare a mash solution suitable for fermentation and distillation. But the basic processes, as well as the desired end result, are the same regardless of the method selected. And that desired result is a distillable solution with an alcohol content of at least 10 percent. In the example that follows, ground corn meal is the basic mash ingredient.

Before distillation can occur, corn must be "broken down"; in other words, the corn's nonfermentable starch must be converted to a sugar that is fermentable. Basically, three things are needed to achieve that breakdown: heat, water and enzymes. The enzymes can be provided by home-prepared malt or they can be purchased from commercial suppliers. Preparing the malt on your own may be cheaper, but more time consuming. I use commercial enzymes because I think it's simpler and easier.

INGREDIENTS FOR MASHING

7	pounds of ground corn meal
2½	gallons of water
4	teaspoons of amylase enzyme (*Canalfa*)
2	teaspoons of gluco-amylase enzyme (*Gasolase*)
1	package (about 2 teaspoons) of distiller's yeast (*saccharomyces cerevisiae*)

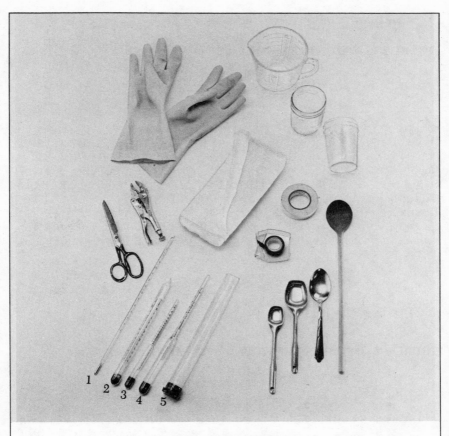

MATERIALS FOR MASHING

1   long-stem laboratory thermometer (optional) (See photo; item 1)

1   floating thermometer (item 2)

1   proof hydrometer (0 to 200 proof) (item 3)

1   triple-scale wine hydrometer (measures sugar content, alcohol potential and specific gravity of mash) (item 4)

1   graduated cylinder (tube for hydrometers) (item 5)

1   long-handled wooden spoon and measuring spoons

1   cheesecloth or burlap bag; citrus bags also work

½   cup of flour for sealing paste

6   mason jars or similar containers

1   vise-grip wrench or pliers

1   pair of rubber gloves

1   roll of tape

    Scissors

FIGURE 2-1. WEIGHTS AND MEASURES

U.S. LIQUID MEASURE

	No. gals.	No. qts.	No. pts.	No. cups	No. oz.	No. T.	No. tsp.	No. gr.	No. lb.
GALLON	1	4	8	16	128	256	768	3789	8
QUART	1/4	1	2	4	32	64	192	947.2	2
PINT	1/8	1/2	1	2	16	32	96	473.6	1
CUP	1/16	1/4	1/2	1	8	16	48	240	1/2
OUNCE	1/128	1/32	1/16	1/8	1	2	6	29.6	1/16
TABLESPOON	1/256	1/64	1/32	1/16	1/2	1	3	15	1/32
TEASPOON	1/768	1/192	1/96	1/48	1/6	1/3	1	5	1/96
GRAMS	—	1/947	1/473	1/240	1/29	1/15	1/5	1	1/473
POUND	1/8	1/2	1	2	16	32	96	473.6	1

FIGURE 2-2. DEGREES CENTIGRADE TO FAHRENHEIT

°C.	°F.	°C.	°F.	°C.	°F.	°C.	°F.	°C.	°F.
0	32.0	23	73.4	46	114.8	68	154.4	90	194.0
1	33.8	24	75.2	47	116.6	69	156.2	91	195.8
2	35.6	25	77.0	48	118.4	70	158.0	92	197.6
3	37.4	26	78.8	49	120.2	71	159.8	93	199.4
4	39.2	27	80.6	50	122.0	72	161.6	94	201.2
5	41.0	28	82.4	51	123.8	73	163.4	95	203.0
6	42.8	29	84.2	52	125.6	74	165.2	96	204.8
7	44.6	30	86.0	53	127.4	75	167.0	97	206.6
8	46.4	31	87.8	54	129.2	76	168.8	98	208.4
9	48.2	32	89.6	55	131.0	77	170.6	99	210.2
10	50.0	33	91.4	56	132.8	78	172.4	100	212.0
11	51.8	34	93.2	57	134.6	79	174.2	101	213.8
12	53.6	35	95.0	58	136.4	80	176.0	102	215.6
13	55.4	36	96.8	59	138.2	81	177.8	103	217.4
14	57.2	37	98.6	60	140.0	82	179.6	104	219.2
15	59.0	38	100.4	61	141.8	83	181.4	105	221.0
16	60.8	39	102.2	62	143.6	84	183.2	106	222.8
17	62.6	40	104.0	63	145.4	85	185.0	107	224.6
18	64.4	41	105.8	64	147.2	86	186.8	108	226.4
19	66.2	42	107.6	65	149.0	87	188.6	109	228.2
20	68.0	43	109.4	66	150.8	88	190.4	110	230.0
21	69.8	44	111.2	67	152.6	89	192.2	111	231.8
22	71.6	45	113.0						

Selecting an Enzyme. You may wonder which enzyme to buy. I have used, with satisfactory results, enzymes from three companies: BIOCON (U.S.), Inc. of Lexington, Kentucky; Novo Laboratories of Wilton, Connecticut; and Miles Laboratories of Elkhart, Indiana. Each company supplies instructions and complete product information. In some cases, these instructions are prepared for large distillers, but they may be adapted for those who want to produce alcohol on a small scale.

BIOCON enzymes are simple for small-scale, home distillers to use because they require no pH adjustment. (The pH is an indication of the acidity of the mash.) Not having to adjust the mash pH simplifies the cooking process. But I'm including sample mash recipes using all three of the companies' products because, for home distilling, all three are about equally good at converting starch to fermentable sugars. BIOCON's recipe follows; Novo and Miles recipes can be found in Appendix B.

The First Steps. The first step, before preparing your mash, is to ensure that all containers, materials and implements are clean. A diluted household chlorine bleach may be used for this purpose. Rinse thoroughly. It is important to reduce bacterial contamination as much as possible. A good second step is to get a notebook or pad of paper and keep track of *every step* along the way. Note ingredients, cooking times, amounts and any other developments you consider important. By doing this, you can isolate problems and learn what works best for you. Finally, if you use a pressure cooker, be sure to read the instructions that come with it.

Preparing Mash With BIOCON Enzymes. Place your pressure cooker or cooking container on a stove or heat source and fill it with 2½ gallons of water. Add two teaspoons of *Canalfa*, a white powder amylase enzyme. Then add the seven pounds of ground corn (cornmeal) and stir the mixture well. Slowly increase the temperature of the mash and continue to stir. *Canalfa* liquefies the mash and helps keep it in a somewhat liquid form, countering the mash's tendency to become lumpy or excessively thick. Even more important, the enzyme begins the breakdown of corn starches into fermentable sugars.

At this stage, your mash may have the consistency of a grainy hot cereal, such as *Wheatena*. Don't be alarmed, though, if it doesn't look exactly like this. It may be thicker, more like a sticky porridge. If the mash becomes so thick that it's difficult to stir, reduce the temperature until the solution returns to a more liquid state. One thing I've discovered in mak-

---

*See Appendix A for a complete list of ingredients and materials, suppliers and manufacturers.

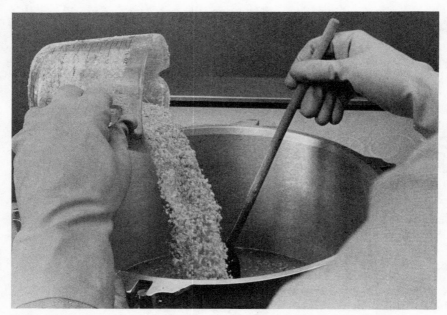

After adding corn to cooker, continue to stir. To prevent excessive thickness, avoid rapid temperature increases. Note that rubber gloves are worn when cooking.

After cooking your mash for about a half an hour, it may have a grainy consistency. If the mash becomes thick or lumpy, reduce temperatures for several minutes until a more liquid state returns. This photo shows about the right consistency.

ing alcohol fuel is that it doesn't pay to rush things. Proceeding at a slow, deliberate pace gets the job done.

After adding the enzyme, continue to stir and gradually increase the mash temperature. Use a floating thermometer to check temperatures. And wear rubber gloves. Plucking a thermometer from a near boiling mash with bare hands is no fun. BIOCON recommends that the mash temperature be increased to about 200° to 212°F. It takes at least an hour to reach that temperature range. Hold at 200° to 212°F. for 15 minutes, while continuing to stir.

Cooling the Mash. Turn off your heat and allow the mixture to cool to 150° to 170° F. Once below 170° F., add two more teaspoons of *Canalfa* and hold for 30 minutes with constant stirring. You can hasten cooling by adding about a gallon of ice water to the mixture, one quart at a time. If additional cooling is needed, fill a plastic bag with ice cubes and place this in the mash. This technique cools the mash without diluting it too much. Another method is to pour the mash out of the hot metal cooking container and into a separate container. (Because pressure cookers retain heat well, it's difficult to cool their contents.) I wouldn't worry if your temperatures are not *exactly* those called for; just be as precise as possible.

Adding Gasolase. Turn off all heat beneath your cooker and cool the mash some more. If you haven't already poured the mash into a separate, clean container, now is a good time to do so, to speed up cooling. As the mash temperature falls below 120° F., add two teaspoons of the enzyme *Gasolase*. This is a "saccharifying enzyme," also known as a *gluco-amylase*, that breaks down the starches further and completes their conversion to fermentable sugars. *Gasolase*, a brown powder, may be added in the form of a paste. Simply mix the powder with a small amount of *cold* water. (Hot water can inactivate the paste.) Forming a paste is not essential, but it does speed up enzyme activity.

Testing for Sugar. After adding the *Gasolase*, test the mash sugar content. Although this test is not absolutely essential, it's a good idea to have the correct sugar content before fermentation commences. For optimum fermentation, the mash sugar level should be about 10 to 15 percent. To determine the sugar content of your mash, use a triple-scale wine hydrometer with a Brix or Balling scale. This scale indicates sugar content by weight.

To make the sugar test, place a piece of cheesecloth or some other suitable straining cloth over a measuring cup or some other small container, such as a mason jar. Hold the cloth in place with an elastic band. Then ladle or pour some of your mash through the cloth and into the container.

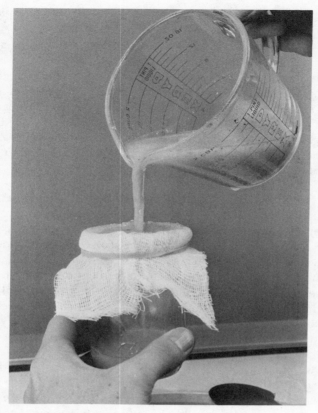

*To strain the mash before testing, simply stretch a
straining cloth across the top of a jar or measuring cup.*

Straining is necessary to achieve accurate readings; suspended solids in
the mash could upset this process.

Reading Your Hydrometer. Pour the contents of the container into the
plastic tube (sometimes called a graduated cylinder) that comes with the
hydrometer. Be careful not to overfill the tube. Leave enough head room
at the top for the liquid that will be displaced when the hydrometer is
placed inside the tube. Lower the hydrometer, weighted end down, into
the liquid. Spin the hydrometer to dislodge air bubbles. Then hold the hydrometer at eye level and read the figures where the liquid surface cuts
across the hydrometer stem. The figures on the Brix scale tell you the percent of sugar in the mash.

An Ideal Sugar Content. Although a good sugar content at this stage,
just before the start of fermentation, is between 10 and 15 percent, I'm

Stem

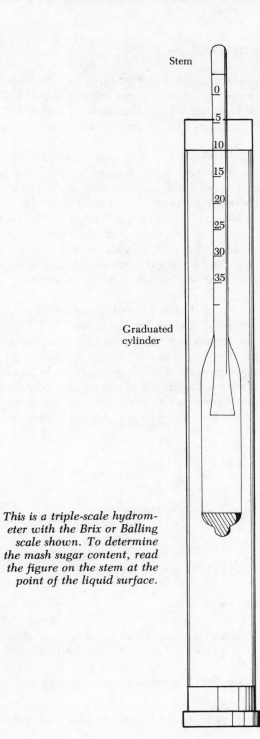

0
5
10
15
20
25
30
35

Graduated
cylinder

*This is a triple-scale hydrom-
eter with the Brix or Balling
scale shown. To determine
the mash sugar content, read
the figure on the stem at the
point of the liquid surface.*

satisfied if my sugar level is even close to 10 percent. More often, it's about 5 to 7 percent. In the unlikely event that your sugar content is above 20 percent, dilute the solution with water until the sugar level is under 15 percent. Excessive sugar concentrations will, in effect, "kill" the yeast to be used for fermentation.

At the opposite extreme, if insufficient sugar is present, the final alcohol yield will be reduced and the time spent fermenting the mash is somewhat wasted. Personally, I would not add expensive sugar to my mash just to bring the sugar content up and to try to recoup my effort to this point. I'd just go ahead and ferment the mash and hope for the best yield possible. However, if you want to add sugar, use this formula: for each full percentage point less than 15 on the Brix scale, add ¼ cup of sugar per gallon of mash. After adding the sugar, recheck another sample in your hydrometer to see if the sugar level is about 10 to 15 percent.

Potential Alcohol. It is also helpful, but not essential, to use your hydrometer to take a reading of potential alcohol. By doing this, you'll know the approximate potential of your mash to produce the desired end product — alcohol fuel. To determine the alcohol content of the mash after fermentation, you must take a reading on the alcohol potential scale before fermentation starts. Make this first reading now and record your potential alcohol percentage. Then, after fermentation, take a second reading. Subtract the second figure from the first to determine the alcohol percentage by volume. For example:

First reading	12 percent
Second reading	4 percent
Alcohol content	8 percent

Fermentation

Continue to cool your mash down to fermentation temperatures (85° to 90°F. is preferable; 70° to 90°F. is acceptable; 60° to 90°F. will work).* Next, convert the sugar in the mash to alcohol. Yeast is the agent that accomplishes this task. In effect, yeast "feeds" on the sugar and breaks it down to alcohol and carbon dioxide. Because yeast works best under anaerobic (oxygen-free) conditions, set up a fermentation vat that allows

---

*BIOCON yeast can tolerate temperatures up to 100°F.; the *optimum* fermentation temperature is 86°F.

the yeast to feed on the sugar, produce alcohol and release carbon dioxide with as little air introduced into the vat as possible. Clean working conditions are especially important here, too, to avoid the introduction of interfering bacteria.

A Plastic Fermenter. A simple way to ferment your mash is to let it stand in a large plastic container with a tightly fitting lid. Puncture the lid and place a fermentation lock into the hole. By using a separate container, you keep your pressure cooker free for the next batch of mash. If you should want to use a cooker as a fermenter, it can be adapted easily for this purpose. Use friction or electrical tape to attach about three or four feet of plastic tubing to the fitting on the lid of the cooker. (I'm referring to the fitting used earlier to attach the copper tubing.) Then extend the opposite end of the plastic tube to a bucket or container of water and submerge it. This provides a way for the carbon dioxide to escape and, at the same time, prevents air from entering the cooker-fermentation vat. A fermentation lock accomplishes the same thing.

Adding Yeast. To get the fermentation going, dump the contents of one packet (about two teaspoons) of distiller's yeast into the mash; then mix the solution slowly and thoroughly. Mix again every two or three hours if possible. If not, mix once or twice per day. Try to do this in a way that allows the least amount of air to come in contact with the mash. Don't be alarmed or disappointed if at first there isn't much reaction. This is normal. Within about a day, the mash will begin to bubble. Based on what I'd read, I expected to see large, gaseous bubbles coming from the bottom and boiling to the surface. However, that's not always the way my mash performs. Often it bubbles like a sparkling ginger ale.

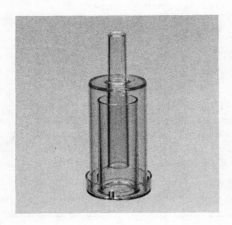

A plastic fermentation lock may be inserted (narrow end first) into the lid of a container to make a fermenter. Fermentation locks may be purchased at wine or beverage-making supply shops.

To make a fermenter from a pressure cooker, tape a piece of plastic tubing to a fitting screwed into the cooker's lid.

A RECAP OF BIOCON PROCEDURES

STEP I. Pour 2½ gallons of water into a 5-gallon container. Add two teaspoons of *Canalfa*. Add seven pounds of ground corn. Mix well. Heat slowly to 200° to 212°F., with constant mixing. Hold for 15 minutes.

STEP II. Cool mash to 150° to 170°F. by adding sufficient ice water or ice pack. Add two teaspoons of *Canalfa*. Mix well. Hold for 30 minutes with constant mixing.

STEP III. Cool mash to 90°F. as fast as possible. Once below 120°F., add two teaspoons of *Gasolase*. Once below 90°F., add two teaspoons of BIOCON's distiller's yeast. Mix well. Let ferment for three to seven days. (Time will be reduced if fermentation temperatures are close to 85° to 90°F.)

BIOCON advises that experienced distillers may reduce *Canalfa*, *Gasolase* and yeast quantities by as much as 50 percent without loss of yield.

If you locate your fermentation vat at a place where temperatures of about 85°F. can be maintained, fermentation should take about three or four days. If mash temperatures are 65° to 70°F., fermentation may take about a week, and possibly longer.

When It's Ready. Old-time moonshiners have reported several different ways to tell when fermentation is complete and the mash is ready for distillation. Usually, when the bubbling stops, this is a sure indication that it's time to start distilling. Sometimes, a cap of foam forms on the surface. It's often about an inch thick at first, but when it sinks or only a few remnants remain on the surface, the mixture is ready to distill. Once bubbling has stopped, don't wait any longer or the mixture will turn to vinegar and smell sour. It's better to begin·distilling a bit early than too late. Sour mash is ruined.

Pre-Distillation

The final step after fermentation and prior to distillation is to *separate the solid and liquid parts of the mash*. This can be done by skimming or siphoning the fluid from the top of the vat and then pouring the remaining mash through a cheesecloth, burlap bag or some other straining cloth, and into a container. (Hold the straining cloth on the lip of the container

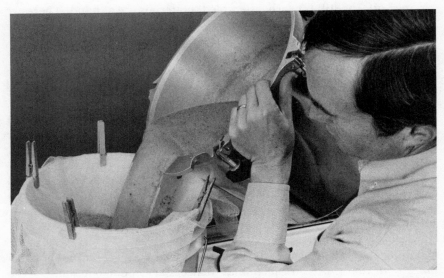

Prior to distillation, pour the mash through a straining cloth to remove solids. A citrus fruit bag, secured with clothespins, works well.

with clothespins.) Some distillers do not make this separation; they distill both the liquid and the alcohol-rich corn residues. I separate the two because I think it's safer and easier to distill just the liquid; this way, there's little chance of excessive foaming or possible fouling of the condenser coil.

Mash Solids: Feed Supplement. Wet or dry, these solids are a high-protein feed supplement. If not fed to livestock or dried within 24 hours, they deteriorate rapidly. A feed should contain no more than 20 percent mash solids. For advice on optimum mash-feed ratios, contact an Extension Service office.

After straining the solids from your mash, pour the liquid, sometimes called *still beer*, back into your pressure cooker. To check the success of your operation so far, take another sample of the still beer and pour it into the hydrometer tube. Examine the scale for alcohol potential, write down the reading and subtract this reading from the one obtained earlier to determine the alcohol content. If your figure is 10 percent or more, you're on the way to a good "run."

Distillation

You are just about ready; mash fermentation is complete; solids are strained off; the still is nearly assembled; materials are at hand. There are just a few remaining items to check before you turn on the stove and fire up the still.

Final Steps. Recheck the coil of copper tubing to make sure it is clean and unobstructed. The easiest way to do this is to blow into one end and place your hand near the other to feel the exiting air.

Having done this, mount the cooling tub or flake stand near the pressure cooker, and attach the free end of the tubing temporarily to the appropriate fitting on the pressure cooker.

Attach the garden hoses to the flake stand fittings, with the hose from your water source entering the fitting near the base of the container. Fill the container with cold water, about 60°F. or less. At this point, you don't need to run the water *through* the container; just let the water stand. Circulation is not yet needed.

Filling the Cooker. Now, return to the pressure cooker. *It should be no more than two-thirds full of liquid*. If necessary, pour out some of the mash so the amount remaining is within the desired range. If the cooker is overfilled, there's a chance the mash will boil over or "puke" during distil-

Before securing the copper tubing to the pressure cooker, blow through it, as demonstrated by author Fred Stetson, to make sure it's not blocked.

The copper tubing may be secured to the lid fitting (compression union) with pliers or a vise grip wrench. Secure the tubing temporarily at first, then tighten it later after heating of the liquid has begun.

29

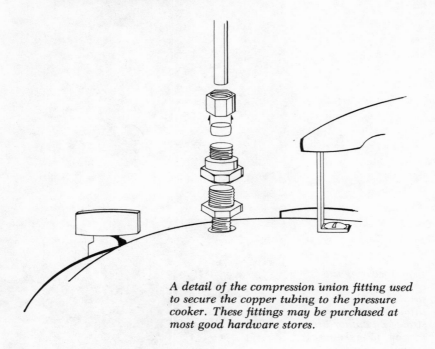

*A detail of the compression union fitting used
to secure the copper tubing to the pressure
cooker. These fittings may be purchased at
most good hardware stores.*

lation. This might plug the copper tubing and, unless the pressure were
released, could cause an explosion. As a precaution, some distillers suggest
that you try running water and then beer through your still, to test the
equipment for leaks and other possible problems. I haven't done this because I've checked for leaks earlier. I'm fully satisfied that the equipment
I've used will work.

Heating the Mash. There's one more step before distillation — simply
place a mason jar or some other small container beneath the "tail" of the
condenser coil. Then turn on the heat beneath the cooker. As the mash begins to warm, detach the tubing, remove the lid and stir to reduce the
chance of scorching. (This is unlikely after solids have been strained off.)
Place your thermometer in the mash and check the temperatures intermittently. When not stirring or checking temperatures, replace the lid to prevent heat from escaping.

It's going to take quite a while to bring three or four gallons of mash up
to near boiling temperatures, perhaps as long as 45 minutes to an hour.
Nonetheless, keep stirring and checking temperatures intermittently. Remember, a nearly pure alcohol solution boils at 173°F., while an alcohol-
water solution like your mash, boils at about 190° to 200°F., depending
on the amount of alcohol.

Keep heating until the mash temperature reaches about 190° to 200°F. Then try to stabilize the mash at a rolling boil. Don't exceed 200°F., if possible. As the mash temperature approaches 212°F., any water in it will vaporize along with the alcohol and dilute your distillate. Ideally, the mash *vapor* temperature should be as close to 173°F. as possible. But I've found that a *mash* temperature of at least 190°F. is needed to get the distillation going.

After the mash temperature has stabilized at about 190° to 200°F., attach the copper tubing to the fitting in the lid of the cooker. You may need pliers or a vise-grip to do this because hot steam may be pushing out of the lid opening. If you use an *All-American* pressure cooker or another cooker with lid wing nuts, screw these down carefully. Screw down two screws on opposites sides of the lid simultaneously, then do the same with another opposite pair. This way, the lid is secured evenly and there's less chance of vapors escaping.

Applying Flour Paste. To prevent vapor loss where the copper tubing is attached to the lid fitting, apply a paste made of a few tablespoonfuls of flour and a small amount of water. Apply this paste directly to the fitting. At first, it will feel and look like a sticky putty, but after about 15 to 30

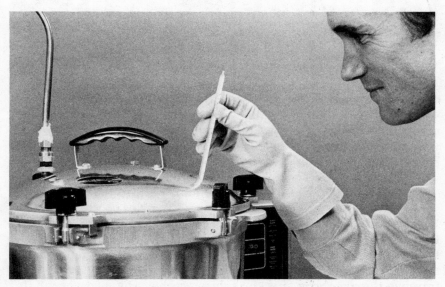

One way to check temperatures without disconnecting the copper tubing is to insert a long-stem thermometer into the lid opening for the cooker's pressure/ temperature gauge. Remember to distinguish between mash and vapor temperatures.

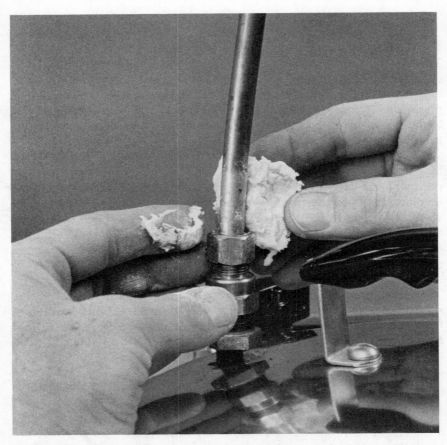

A flour-water paste should be applied liberally to the copper tubing where it connects the compression union fitting.

minutes, it will harden and seal the fitting. It may take a few tries to make your paste harden effectively. The trick is to get the paste on the fitting before too much pressure or heat builds up.

Telltale Signs. Pure alcohol vapors are invisible, so you probably will see little or no sign of their escaping from a fitting or poorly secured lid. But there are some signs to look for that indicate the onset of distillation. For one, the copper tubing extending from the cooker to the flake stand begins to get quite hot. If you take a potholder or rag and wrap it around this stretch of tubing, you actually may be able to feel the alcohol vapors traveling through. Soon after this tubing gets hot, you might hear a tiny puff or muffled gurgle near the end of the coil. Watch. The first trickles of liquid alcohol are about to push out. Turn on your faucet so water circulates in the flake stand.

The First "Shots." Soon the trickle turns to a tiny, slow stream, the thickness of a thin pencil. This stream may be intermittent, but, if the mash temperature is reasonably stable, the stream remains steady. The first alcohol "shots" may be as much as 140 proof or 70 percent alcohol. (Alcohol with 100 proof is 50 percent alcohol and 50 percent water.) After the first pint or so, the proof declines. Use a *proof* hydrometer to determine the proof of your distillate. This instrument is not the same as the wine hydrometer used earlier to measure alcohol *potential*. The proof hydrometer measures actual alcohol content.

As the mason jars begin to fill, you'll notice that the alcohol is very clear. If there was any scorching of mash, the distillate may have a tiny bit of sediment in it. But, generally, the alcohol looks crystal clear, smells like a strong, raw alcohol and, at least to me, feels like slightly contaminated water.

Alcohol exits from your condenser coil at varying rates; usually, when temperatures are steady, the flow is a thin stream, slightly faster than a trickle.

Test for Proof. Usually, I allow about a half-pint of alcohol to accumulate in a jar, then I take the jar away and replace it with an empty one. Next, I test the proof of each jar and label it; this way, I can keep track of the strength of the alcohol.

When the proof declines below about 100, I turn off the stove. The remaining mash can be added, at the start, to the next batch of mash to be cooked. This step is sometimes called "sloppin' back." It's a way to use the alcohol-rich liquid as either a partial or a full replacement for the water initially added to the corn when mash is cooked for the first time.

Rerunning Alcohol. Following the steps outlined in this chapter, you should be able to produce up to a half gallon of low-proof alcohol during a single run lasting 90 to 120 minutes. This alcohol will probably have a proof of only 100 to 150. To strengthen the liquid, you must rerun it for about 45 to 60 minutes. This rerun will increase the proof of most of the first-stage distillate to about 160 to 170. Most experts agree that a minimum of 160-proof alcohol fuel is needed to operate a combustion engine. To save time, you might accumulate several quarts of low-proof alcohol and then use this combined batch for your second run.

Do not drink the alcohol you produce. It may be poisonous.

CHAPTER 3

Building and
Using a 55-Gallon Still

In simplest terms, my 55-gallon still is a "scaled-up" version of the 5-gallon still described in Chapter 2. Operating the larger still is a bit more complicated than operating the smaller one. While the larger version has greater capacity, the amount of working time and effort is also greater. The larger still can produce up to 2½ gallons of alcohol per bushel of corn processed. (A bushel of corn weighs 56 pounds.) The still can be made from readily available, inexpensive parts. Black steel was selected for the distillation column. More durable and superior metals, such as copper and stainless steel, could have been used, but these are also more expensive.

One point to keep in mind: Individuals, universities, companies and corporations of all sizes are rapidly developing new and ever-more-efficient alcohol distilleries and equipment. If you intend to produce large quantities of alcohol on a continuous basis, read the description of large stills in Chapters 5 and 6 and refer to the Appendices. If you anticipate investing a lot of time and money producing alcohol, seek well-qualified, professional assistance. Review the literature from various companies and *examine their stills in operation.* Before making your investment, ensure that the products you buy will produce usable, high-proof, fuel-grade alcohol. State universities and Extension Service offices in many parts of the country, especially the rural states, are able to help assess alcohol fuel-making equipment.

How the 55-Gallon Still Works

Briefly, here are the steps you will follow when working with the 55-gallon still: Mix a corn-water mash solution in a 55-gallon drum heated by a wood fire. The firebox is a half drum welded to the base of the

Mash cooking is underway in this wood-fired still, fabricated from 55-gallon drums and readily available parts. The distillation column projects from the drum at left; the copper condenser coil is inside the drum at right.

55-gallon drum. Cook the mash, treat it with enzymes, cool it and drain it off into plastic buckets. Next, treat the mash again with enzymes, then yeasts and allow it to ferment for about three to six days.

After fermenting the mash, stretch a piece of straining cloth across the top of a container and pour the mash through it to separate solids from the alcohol-rich liquid. Pour this liquid back into the 55-gallon cooking drum. Reignite the fire, heating the liquid enough so that alcohol vapors pass up into the distillation column. The vapor becomes increasingly alcohol rich as it ascends and, ideally, most of the water remains in the column or falls back into the cooking vat. The alcohol vapors continue on to a copper coil where they condense and pass out through the tail of the coil as liquid alcohol. If the alcohol proof is below 160, the alcohol must be re-run through the still.

BUILDING THE 55-GALLON STILL

The 55-gallon still has four basic parts:

- A firebox, made of half a 55-gallon drum and fitted with a furnace clean-out door and a flue.

- A 55-gallon drum used as a combination cooking and distillation vat and mounted on top of the firebox.

- A three-inch, black steel distillation column, packed with glass marbles.

- A copper coil, mounted inside a 55-gallon cooling drum, that acts as a condenser.

Fabricating a workable still from these parts requires a fair amount of welding and metal work. If you are experienced with welding and metal-cutting tools, making this still will not be difficult. If you're inexperienced, I'd suggest seeking some help. I did. And it saved a tremendous amount of time. You can find materials needed for this still at dumps or purchase them from plumbing, heating, welding or materials suppliers. New, they'll cost about $250 to $350. But used parts may be purchased at little or no cost.

FIREBOX MATERIALS

1 55-gallon drum

1 furnace clean-out door

1 length of angle iron, for door frame ($\frac{1}{8}$" × 1" × 3" × 57$\frac{3}{8}$")*

4 $\frac{3}{8}$"-hex head bolts ($\frac{3}{8}$" bolt, $\frac{9}{16}$" head)

4 $\frac{3}{8}$"-hex head nuts

1 4"-long piece of stovepipe collar to accommodate 6" diameter stovepipe lengths and elbows.

As many stovepipe lengths and elbows as necessary for flue (24 gauge is satisfactory)

---

*Metal shaped or bent to form angle iron may be substituted.

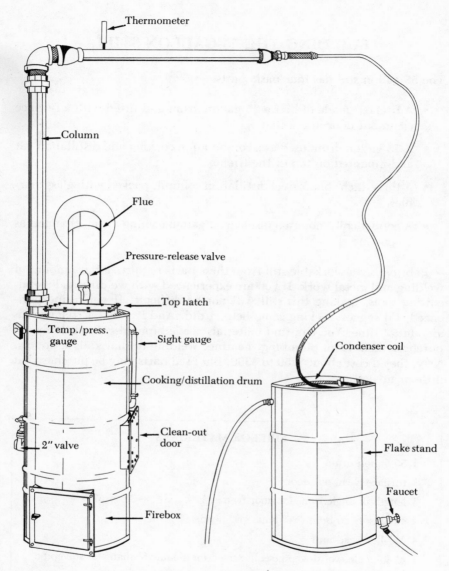

Thermometer

Column

Flue

Pressure-release valve

Top hatch

Temp./press.
gauge

Sight gauge

Cooking/distillation drum

Condenser coil

Clean-out
door

2″ valve

Flake stand

Faucet

Firebox

This is the still when completed and ready to operate. Mash is cooked and, after fermentation, distilled in the 55-gallon drum at left. Alcohol-rich vapors escape through the column, then move to the condenser coil where they reform as alcohol fuel.

Before you get started, examine the lists of materials on pages 37, 45, and 54 and assemble as many of these as possible. Notice you'll need a total of three 55-gallon drums. Select them carefully. Make sure they are clean and free of chemicals or possibly flammable materials. Remove all caps from their bungs to ensure there is no chance of a fume build-up or excessive pressure. Select three identical drums if possible; be sure, at least, that two of the three drums have the same diameter.

The Firebox

Using a saber saw, cut in half one of the two 55-gallon drums with identical diameters. If you have a metal band that will wrap around the drum, this might be used as a guide for marking where the drum is to be cut. Cut the drum cleanly and evenly about 18 inches from the bottom.

A metal band is a handy device for ensuring that the drum is marked and cut evenly.

The remaining edge will later be welded to the base of the second 55-gallon drum, the one with the identical diameter. The bottom half of the drum just cut will become the firebox.

Firebox Door. As with many other phases of still building, there is no one best way to fabricate a firebox door. It may be made of separate pieces of flat metal and angle iron. Or, you can acquire a small furnace clean-out door. Such a door comes with its own frame, but you'll need a second frame to attach the door to the drum. The inside dimensions of my door frame, or the frame opening, are 12 × 12 inches. If your clean-out door is a different size, simply adjust all other dimensions as necessary.

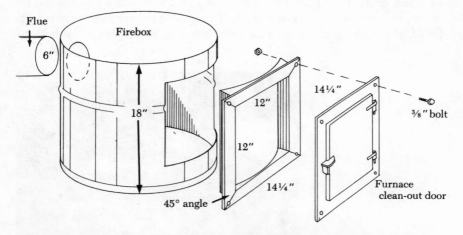

Firebox, firebox door frame and clean-out door, which has its own frame.

Take your piece of angle iron (⅛ × 1 × 3 × 57⅜ inches) and cut it into four pieces 14¼ inches long. Cut the ends so they'll fit together at 45 degree angles.* These pieces may be welded together, as they are, with straight edges to form the second frame. Or, the top and bottom pieces may be cut so they'll fit neatly with the rounded edge of the firebox drum. To do this, first make a template of cardboard or thin sheet metal. This template should be marked and cut so one edge is a curve, corresponding

---

*It is not necessary that the angle iron be cut to a 45-degree angle. This is merely a neater way to make the welds. If the piece of angle iron is long enough, it may be cut into four pieces with square ends suitable for welding into a frame. In this case, two pieces should be about 16 inches long and two pieces should be about 14 inches long. Remember, in making the frames for the firebox door (and the 55-gallon drum's clean-out door and top hatch), it is not necessary to duplicate my dimensions precisely.

Cut a template with a rounded edge, if you want to make the edges of the door frame round so they'll conform to the circumference of the drum.

Once the four pieces of the door frame are welded together, buff the frame to remove rough edges. Use the frame opening as a guide for cutting the door hole in the firebox.

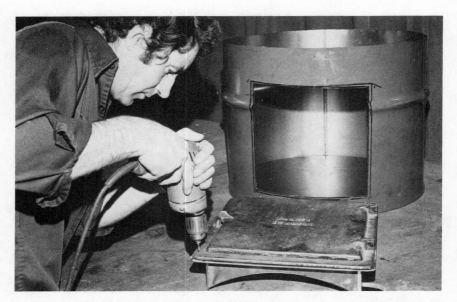

Place the door with its own frame on top of the second frame (fabricated from four pieces of angle iron), and then drill four 7/16-inch holes through both frames, one hole in each corner.

Firebox with frame welded to drum. Note that notches have been cut in frame edges, so the edges may be welded flush to firebox drum at all points.

to the circumference of the drum. Place the template on the two pieces that will become the top and bottom of the frame. Use the rounded edge of the template to mark a curve on the two pieces of angle iron. Then cut a curved edge in the angle iron to create an edge nearly identical to the curved surface of the drum.

Once the four pieces are welded together, buff the frame to remove rough edges. Place the frame so its bottom is about two inches from the bottom of the firebox drum. Using the frame opening (which should be about 12 × 12 inches) as a guide, mark and then cut a hole in the firebox drum of the same size.

Place and clamp the door with its own frame on top of the second frame. Drill four $7/16$-inch holes through both frames, one hole in each corner. Finally, weld the second frame to the firebox drum. And, at your convenience, attach the door and its frame to the second frame with $3/8$-inch hex head bolts and nuts.

Stovepipe Collar. The last step is to make a stovepipe collar from a four-inch-long piece of six-inch diameter pipe. Fit and weld this piece into a six-inch diameter hole cut in the back of the firebox, approximately opposite the firebox door and about one to two inches below the top edge. (The stovepipe diameter probably could be reduced to four inches, if the flue needed for adequate draft consisted of only a short length of pipe with no more than one elbow.)

The stovepipe and elbows shouldn't be installed until later, not until the cooking/distillation drum is complete and in position, with the col-

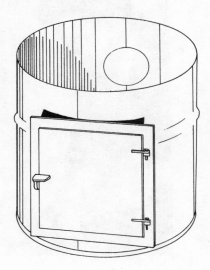

Firebox with door attached. A six-inch hole is cut in the rear of the firebox for the stovepipe collar and flue.

umn and copper cooling tubing attached. As a final step prior to lighting your first fire in the box, place about two inches of sand over the bottom of the firebox. This sand will protect it and increase its lifespan. Six to eight bricks may also be arranged on the sand, to form improvised andirons. These andirons are not absolutely necessary, but they will bring the flames and heat of your fire closer to the mash.

Cooking and Distillation Drum

After completing the firebox, put it aside and prepare to work on the 55-gallon cooking/distillation drum. The general sequence of steps will be:

- Cut two square openings in the drum and then fabricate door frames and plates to cover the openings.

- Cut several small holes in the drum for a gate valve, pressure-release valve, temperature-pressure gauge, sight gauge and distillation column.

- Weld most of these parts (or the adapters into which they are screwed) to the drum.

- Weld the 55-gallon drum to the firebox.

The first hole to be cut in the 55-gallon cooking and distillation drum is a 12 × 12-inch hole in the top surface.

Top Hatch. Begin by cutting a square opening, measuring 12 × 12 inches, for a hatch on the drum top. Later, you'll pour corn and stir your mash through this opening. Next, cut your length of angle iron (⅛ × 1 × 2 × 57⅜ inches) into two 14-inch pieces and two 14½-inch pieces. (Cut ends to 45-degree angles.) Weld these four pieces together to form a frame. But don't weld the frame to the top of the drum yet.

COOKING AND DISTILLATION
DRUM MATERIALS

1   3′ × 5′ piece of foil-backed insulation (fiberglass batts)

1   55-gallon drum with diameter equal to firebox drum

1   2″ gate valve

1   2″ nipple, 4″ long

1   length of angle iron, measuring ⅛″ × 1″ × 2″ × 57⅜″

1   piece of flat, sheet steel (mild), measuring ⅛″ × 14½″ × 14″

2   ⅜″ nipples, each 2″ long

2   ⅜″, 90-degree elbows

1   ½″ diameter piece of *Tygon* tubing, 15 inches long

2   hose clamps

1   ½″ coupling

1   combination gauge, with a 0° to 212°F. temperature range and a 0 to 30 pounds pressure range

1   30-pound pressure release valve (hot water tank type)

1   ¾″ nipple (black steel pipe), 6″ long

1   ¾″ nipple (black steel pipe), 3″ long

1   ¾″, 90-degree elbow (black steel)

1   ¾″ nipple (black steel pipe), 4′ long

1   length of angle iron, measuring ⅛″ × 1″ × 2½″ × 49⅜″

1   piece of flat sheet steel (mild), measuring 1″ × 12¼″ × 12¼″

24   ⅜″-hex head bolts

24   ⅜″-hex head nuts

1   piece rubber gasket, 14½″ × 14″

1   piece rubber gasket, 12¼″ × 12¼″

1   container of "pipe dope" (plumbing sealant)

First, cut a piece of ⅛-inch flat sheet steel that will cover the top of the frame. This should measure ⅛ × 14½ × 14 inches. But double check the exact size of your frame, then cut your piece of flat steel so it covers the frame neatly.

Mark 12 spots, spaced evenly along the edges of the sheet steel, for ⁷⁄₁₆-inch holes. These holes will accommodate ⅜-inch hex head bolts. Clamp the flat sheet on top of the frame and drill 12 holes (⁷⁄₁₆ inch) through both the sheet and the frame. Place a ⅜-inch hex head bolt through each of the holes; secure each bolt with a ⅜-inch nut. Turn the frame upside down and weld each nut to the underside of the frame. Once the welds are set, remove the bolts. It's not absolutely essential that the nuts be welded to the underside of the frame, but when it's time to take the cover plate on and off, you'll be glad they are welded in place.

Weld the frame to the top of the drum. Finally, put a 14½ × 14-inch piece of gasket material over the frame, cover it with the piece of flat sheet steel, make sure the pieces are aligned over the frame, then drill 12

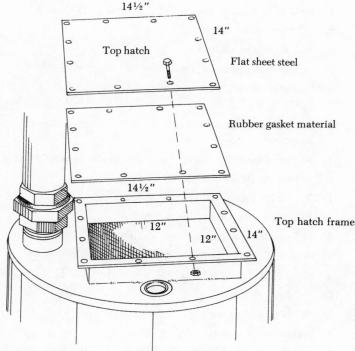

Drill holes through sheet steel and gasket. The two pieces are secured to the frame with ⅜-inch hex head bolts and nuts. Weld nuts to the underside of frame.

Completed top hatch showing ⅜-inch hex head bolts. It's a good idea to place a reference mark on one edge of the steel plate, so the plate can be easily and correctly reoriented each time it is placed on top of the hatch frame.

holes through the gasket material. Drilling rubber may be difficult. You might try sharpening a section of ⅜-inch copper pipe and using it as a punch. When you're ready to use the drum for distilling, use the ⅜-inch hex head bolts and nuts to secure the sheet steel and gasket tightly in place.

Clean-out Door. As its name suggests, the clean-out door provides an easy way to clear the drum of mash after cooking. Making the door and its frame is similar to the fabrication of the top hatch and its frame.

To start, cut the second piece of angle iron (⅛ × 1 × 2½ × 49⅜ inches) into four pieces each 12½ inches long. Two of the pieces, for the top and bottom of the frame, may be used as is or their edges (the edges to be welded to the drum) may be cut to form a curve, conforming with the circumference of the drum. If the edges are cut, use the same steps outlined for the firebox frame.

Next, cut the ends of the four pieces to 45-degree angles and weld the four pieces together to form your clean-out door frame. Then hold the frame right at the base of the 55-gallon drum, so the clean-out door will be positioned as shown on page 48. Outline the inside of the frame on the drum with a heavy, felt-tipped pen. Then cut a square piece out of the drum—a piece that will correspond with the inside dimensions of the clean-out door frame.

Finally, cut a 12¼ × 12¼-inch piece of ⅛-inch-thick flat, mild steel.

47

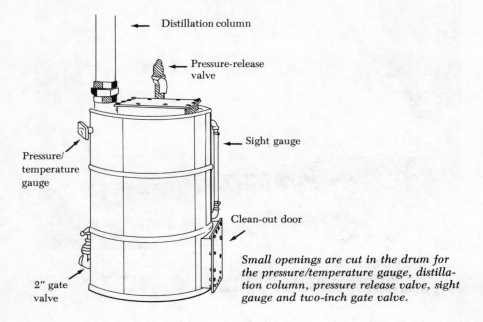

← Distillation column

← Pressure-release valve

← Sight gauge

Pressure/ temperature gauge

Clean-out door

2" gate valve

Small openings are cut in the drum for the pressure/temperature gauge, distillation column, pressure release valve, sight gauge and two-inch gate valve.

This piece will be the drum clean-out door plate. It won't be hinged to the frame; rather, it will be bolted in place over a 12¼ × 12¼-inch piece of rubber gasket material, to form a tight seal.

Clamp the flat steel in place over the frame with C-clamps. Mark places for 12 evenly spaced, ⁷⁄₁₆-inch holes along and through the edges of the plate and frame. Drill the holes. Insert a ⅜-inch hex head bolt through each hole; secure each bolt with a ⅜-inch hex nut. Weld the nuts to the underside of the frame. Once the welds are set, remove the bolts.

Cut a 12¼ × 12¼-inch piece of gasket material, with a thickness of about ⅛-inch. Align the piece on the frame and drill or puncture 12 holes through it, corresponding with the holes in the frame and door plate.

The last step is to weld the frame to the 55-gallon drum. Use care to ensure that there are no leaks. (Later, after all the parts of the drum have been welded, fill the drum with water and check for possible leaks.)

Smaller Drum Openings. There are five small holes to be cut for remaining fixtures on the 55-gallon drum. These holes may be started with a drill and then enlarged with a saber saw. Seal all threaded joints with "pipe dope" (any commercial plumbing sealant). The five holes are:

• A two-inch diameter hole for the two-inch gate valve, located at the base of the drum and positioned as shown in the illustration above.

Cut this hole and then weld the gate valve nipple in place. Although the valve may seem a bit oversized, the extra space is necessary to permit easy drainage of the mash after cooking. Smaller valves might plug or foul with the corn-water mixture.

- Two ½-inch holes are cut in the upper third of the drum above the clean-out door. Cut one hole about 15 inches below the other, with the top hole about 1 to 2 inches from the drum top. These holes are for two elbows. The nearly clear, plastic *Tygon* tubing will be your sight gauge. Weld elbows to the drum, screw one short ⅜-inch nipple (2 inches long) into each elbow, then secure the tubing to the nipples with hose clamps.

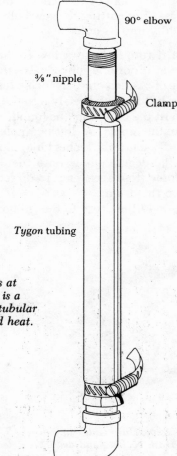

90° elbow

⅜" nipple

Clamp

Tygon tubing

Two holes are cut for the 90° elbows at either end of the sight gauge. Tygon is a nearly clear material, suitable for a tubular gauge that must be able to withstand heat.

- Cut a one-inch hole, about 3 inches from the top of the drum for the temperature/pressure gauge (for position, see page 48). Weld a coupling in this hole for the gauge. The gauge not only warns of possible pressure build-ups, it provides an indication of *vapor* temperatures near the base of the distillation column. Remember, during distillation, *vapor*, not mash, temperatures are critical. Mash is heated, as necessary, to achieve the optimum vapor temperatures for distillation. (To avoid possibly damaging the gauge, don't screw it into the coupling until after the drum has been welded to the firebox.)

- Cut a 3-inch hole in the top surface of the drum, to the left of the hatch, for the 3-inch black steel distillation column. This hole must be cut because most large bung holes on the tops of 55-gallon drums are smaller than three inches in diameter. Mine, for example, is 2¾ inches. After cutting the new hole, replace the large bung and screw it tightly into its hole.

The last fixture to be attached to the drum is the 30-pound pressure-release valve. If possible, buy a valve that fits on a ¾-inch threaded nipple (a piece of ¾-inch pipe, six inches long). If the diameter is ¾ inch, the nipple may be screwed directly into the smaller bung on the drum top. If a nipple to fit the bung can't be found, cut a new hole on the drum to accommodate the nipple. Weld the nipple to the drum; screw on the valve; screw the ¾-inch (three inches long) nipple into the valve exit hole; screw a ¾-inch, 90-degree elbow onto the 3-inch nipple; and add the third ¾-inch nipple (four feet long) to direct any steam downward, should it escape from the drum.

Finally, weld the drum to the firebox.

A pressure/temperature gauge like this is ideal for mash cooking and distilling.

This is a typical 30-pound pressure release valve with accompanying nipple at left.

Fabricating a 55-gallon still is not difficult for experienced welders. Those who are inexperienced should seek professional help.

Distillation Column

The column is a series of pipes leading from the top of the 55-gallon drum to the condenser. A vertical, 56-inch section of three-inch, black steel pipe is packed with marbles. These marbles help to separate the alcohol vapors from the ascending alcohol/water "steam." If conditions are ideal, the alcohol vapors continue their ascent up the pipe, while most of the water condenses and returns to the drum. To achieve this kind of separation, temperatures must be controlled as carefully as possible. Two thermometers, one (the combination gauge) near the top of the drum and one near the top of the column, enable temperature monitoring.

To make the column, follow these steps:

- Weld a 3-inch-diameter section of black steel pipe (nipple) to the top of the 55-gallon drum at the appropriate hole. The section should be about three inches long and threaded at the upper end.

- Screw half of a 3-inch threaded union onto the threaded end of the 3-inch pipe. The union will join the 3-inch-long section to the 56-inch section. (All threaded connections, except the unions, should be sealed with "pipe dope" or a commercial sealant to ensure that vapors will not escape.)

- Weld a 3-inch-diameter disk of perforated steel to the inside of the

51

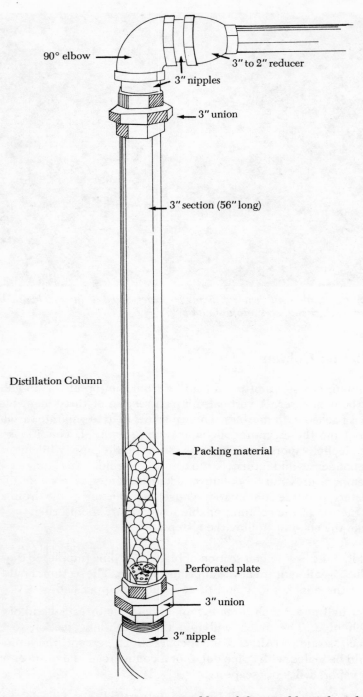

90° elbow

3" to 2" reducer

3" nipples

3" union

3" section (56" long)

Distillation Column

Packing material

Perforated plate

3" union

3" nipple

The three-inch unions permit easy assembly and disassembly of the column.

base or bottom of the 56-inch section. This disk supports the marbles.

- Screw the second half of the threaded union onto the threaded end of the 56-inch-long section of 3-inch pipe.

- From this point forward, take your 56-inch-long pipe and add pieces to it until the column is complete. Then, using the 3-inch union, attach the completed unit to the drum. Find at least one other friend to help you do this. It's a cumbersome job and it requires a large wrench. I had to borrow a wrench from a plumbing supplier.

- The next piece to add is another 3-inch union on top of the 56-inch section. (This union is not absolutely essential. The 56-inch section could be attached directly to the next major piece — a 90-degree elbow that directs the vapor toward the condenser. But the union does enable easy disassembly of the distillation column.)

- If you do add this second union, screw a 3-inch nipple (four inches long) into the top of it. This small pipe section connects the union to the 90-degree elbow.

- Screw a 3-inch, black steel, 90-degree elbow onto the 4-inch long pipe.

Close up of the 3-inch perforated plate welded to the base of the 56-inch-long pipe.

A detail of the upper column, showing union, elbow, nipples and thermometer.

- Screw another 3-inch nipple (four inches long) into the open end of the elbow. This pipe section joins the elbow to a reducer. The reduction is from 3 to 2 inches in diameter.

- After screwing the reducer in place, take the 2-inch black steel nipple (24 inches long), threaded at both ends, and drill a hole in its upper side for a ¾-inch coupling. Weld the coupling to the nipple. When ready to distill, screw the 0° to 212°F. thermometer into the coupling.

- Screw the 2-inch-diameter black steel nipple (24 inches long) into the reducer, and then screw a second reducer onto the open end. The second reduction is from 2 inches to 1.

- Screw the 1-inch nipple (3 inches long) into the 2-to-1-inch reducer. The 1-inch nipple should have half of a 1-inch union screwed onto its open end.

DISTILLATION COLUMN MATERIALS

1	3″ black steel nipple (a section of 3″-diameter pipe; should be 3″ long)
2	3″ unions
1	3″ black steel pipe, 56″ long
1	3″ disk of perforated steel
1	3″ black steel, 90-degree elbow
500	marbles; electrical insulators are another suitable column packing material
2	3″ nipples, 4″ long
1	0° to 212° F. thermometer that fits into ¾″ coupling
1	¾″ coupling
1	3″ to 2″ reducer
1	2″ black steel nipple, 24″ long
1	2″ to 1″ reducer
1	2″ nipple, 2″ long
1	1″ nipple, 3″ long
1	1″ union

Parts connecting the column to the copper condenser tubing are, from left, two-inch nipple, second reducer (2 inches to 1 inch), one-inch nipple, one-inch union, one-inch nipple, copper female adapter and copper tubing.

At this point, the column may be left as it is, and you can turn your attention to the cooling tub. After the tub's copper coil is welded in place, it will join the distillation column at the one-inch union. Discussion of this connection continues later.

Cooling Tub or "Flake Stand"

The cooling tub, which moonshiners refer to as a *flake stand*, is nothing more than a 60-foot-long piece of ¾-inch copper, coiled so it fits inside the third 55-gallon drum. Vapors from the distillation column move to the coil, where they condense into alcohol distillate. Cold water cools the coil. The water enters and exits the drum through garden hoses.

To make your tub, cut out one end and then cut three holes in the drum:

- One hole, about five inches from the top of the drum should be the right size for a coupling that will accommodate a common garden-hose fitting. In some cases, the garden-hose fitting may be suitable for welding directly to the drum.

- Cut a second hole near the base of the drum, where a nipple for a faucet will be welded. This is where the cold water will enter the drum. The faucet enables water control at the drum, rather than at the water source.

- Cut a third hole with the same diameter as the copper tubing (¾ inch in this case), near the drum base. This will be the exit hole for the tail of the copper coil.

After cutting these holes, weld the hose fittings to the drum. (If they are copper, soldering, rather than welding, is necessary.)

Next, carefully wrap the copper tubing around a round object so it forms a reasonably neat coil, small enough in height and diameter to fit inside the 55-gallon drum. Use an appropriate-size log or barrel for this task.

Leave about six to eight feet of the top of the coil free and approximately straight. This will be used to connect the coil to the distillation column. Also leave at least 10 inches straight at the bottom. This part of the coil will exit through the side of the drum.

COOLING TUB
OR "FLAKE STAND" MATERIALS

1   55-gallon drum

60′ ¾″ copper tubing*

1   copper female adapter, ¾″ on one end; 1″ and threaded on the other

6   pieces of strip metal, each about 1″ wide and 6″ long

3   pieces of strip metal, each about 1″ wide and 3 feet long

1   male garden-hose fitting

1   coupling for garden-hose fitting

1   faucet

1   nipple for faucet

2   lengths of garden hose

1   container of about 1 to 2 gallons capacity

3   1-gallon plastic milk jugs

2   small plastic food containers

---

*Any length, from 30 to 60 feet, with a diameter of ½ to ¾ inch, is satisfactory. Copper tubing is often sold in 60-foot lengths. If diameter other than ¾ inch is used, adjust size of copper adapter as necessary.

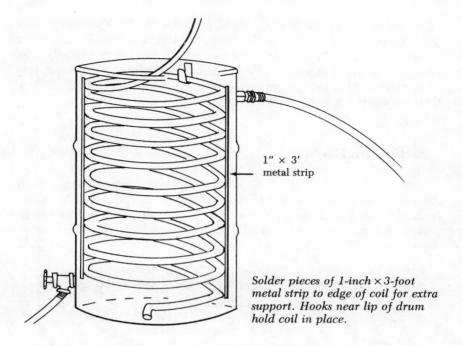

1" × 3'
metal strip

*Solder pieces of 1-inch × 3-foot
metal strip to edge of coil for extra
support. Hooks near lip of drum
hold coil in place.*

Take one of the three pieces of one-inch by three-foot strip metal and
line it up vertically, along the edge of the copper coil. Solder the strip so it
braces the coil and keeps each turn reasonably well spaced. The coil may
tend to collapse like an accordian without this extra support. Repeat these
steps for all three strips.

Hang the coil inside the drum with six evenly spaced "hooks,"
fabricated from pieces of one-inch by six-inch strip metal. Bend the strips
in the shape of a hook, weld them to the sides of the drum, then lower the
coil onto the hooks.

Solder the coil tail so that it projects several inches through the exit
hole. It's a good idea to let the tail come through at least five or six inches.
That'll give you more room to collect alcohol, when it starts trickling
through the coil.

Solder the ¾-inch copper female adapter to the free, upper end of the
copper tubing. The ¾-inch end of the fitting obviously fits over the
¾-inch copper, while a 1-inch nipple, about five inches long and
threaded at both ends, should be screwed into the 1-inch end. Screw half
of the 1-inch union to the open end of the 1-inch nipple. This will be the
point where the copper tubing will connect to the distillation column.

With the welding and soldering complete, test your cooling tub and the
cooking-distillation drum for leaks. Fill both with water, and watch for

signs of drips or sprays. Minor, tiny leaks in the cooking-distillation drum may not be a problem. I found a leak near the base of the frame to my drum clean-out door. But, the first time I used the drum to cook, the mash apparently sealed the leak. Leaks in the cooling tub are a different story. They should be sealed by solder. If water leaks at the tail of the coil, there's a chance it will mix with — and dilute — your alcohol.

Setting Up the Still

To set your still up easily, first separate it into its three basic parts: the 55-gallon cooking-distillation drum; the column; and the cooling tub. Place the cooking-distillation drum on a support about a foot off the ground. I used two concrete blocks, placed side by side, for this purpose.

The 55-gallon drums need not be elevated, but the still is easier to use if they are.

The extra height enables you to feed the firebox without excessive bending or leaning.

With the 55-gallon drum in place, install the firebox flue. If you're working outside, a simple elbow connecting the firebox to a straight, vertical stovepipe should be satisfactory. My stove is set up in a garage, so I used two elbows to extend the pipe up and then horizontally through a metal plate installed in a window. Once outside, another elbow turns the pipe 90 degrees upward. The top of the pipe is protected with a small weather cap.

Next, position the cooling tub so the pipe extending from the distillation column can be easily attached to the straight end of the copper condenser coil. It may be necessary to elevate the drum to make this attachment. I elevated my cooling drum by placing it on three large log sections, each 15 to 20 inches in diameter. Finally, with the cooking and cooling drums in place, blow through the copper coil, and then attach the straight end to the pipe extending from the column.

Remember to place a layer of sand across the bottom of your firebox, and use bricks as andirons to elevate the fire.

OPERATING A 55-GALLON STILL

You can use your 55-gallon still indoors or outdoors. Mine is set on a level, concrete floor in my garage. A stovepipe flue nine or ten feet long vents smoke from the firebox to the outside. If you decide to operate your still in a garage or shop, select a well-ventilated location with plenty of clearance from combustible materials. While I am cooking mash and distilling, I leave the garage windows and doors open. I would not recommend setting up the still in the basement of a house.

INGREDIENTS FOR MASHING

1	bushel (56 pounds) of ground corn meal
20-25	gallons of water
9	teaspoons of amylase enzyme (*Canalfa*)*
6	teaspoons of gluco-amylase (*Gasolase*)*
12	teaspoons of BIOCON distiller's yeast*

*With experience, you may reduce these amounts as much as 50 percent, according to the manufacturer, BIOCON (U.S.) of Lexington, Kentucky.

If you've reached this point without building a smaller still, review the steps for mashing and distilling with a five-gallon unit as described in Chapter 2. Many of the steps that follow are explained in more detail in the earlier chapter. And even if you haven't built a small, five-gallon still, you could practice all the *mashing* steps using any suitable saucepan or pot. A still is not needed. Practicing on a small scale makes sense, before you dump large quantities of water and ground corn into a 55-gallon drum.

Mashing

Cooking mash in a 55-gallon drum is similar to mashing with a 5-gallon pressure cooker. Many of the preliminaries are the same; the desired end result is the same. What we're looking for is a mash that, after cooking,

---

MATERIALS FOR MASHING

1   triple-scale wine hydrometer*

1   proof hydrometer (0 to 200 proof)*

1   four-foot length of wood

1   floating thermometer (0° to 212°F.)*

1   burlap bag or straining cloth

6   clothespins

1   measuring cup and measuring spoons

5   small containers (1 quart or less)

6-8 plastic five-gallon containers

1   vise-grip wrench or pliers

1    $\frac{9}{16}$-inch wrench

1   pair of heavy-duty rubber gloves

     dry, split firewood (hardwoods such as maple or oak are preferable)

     rags or old towels

1   spool of wire

1   sheet of fiberglass insulation

1   notebook (for a suggested checklist, see Appendix D)

---

*If you plan to make large quantities of fuel, buy two of each of these items. They are easy to lose or break.

has a concentration of fermentable sugar. Through fermentation the sugar is converted to alcohol; and through distillation the alcohol is separated from water.

Materials and Equipment. To start, make sure you have all necessary materials and equipment. You will need most of the materials and equipment, and all of the ingredients, in larger quantities, used for mashing with a five-gallon still. (See complete lists on pages 59 and 60.) These are a few of the special materials and tools that are either essential or helpful:

- dry firewood, split to small pieces (hardwoods no more than 5 inches in diameter and 18 inches long)
- a sheet of fiberglass insulation batting, large enough to cover the cooking-distillation drum
- a 9/16-inch wrench for tightening bolts on the clean-out door
- rags or old towels
- a floating thermometer (0° to 212°F.) attached to a wire so that it can be lowered into the drum and retrieved

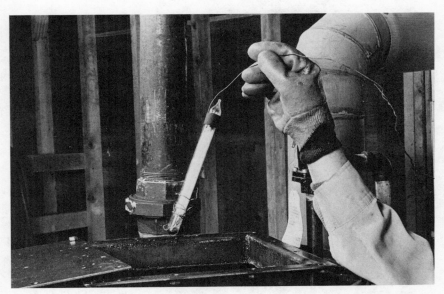

Attach a wire to thermometer so it may be retrieved easily from the cooking drum.

- a length of wood four or five feet long for stirring the mash

- six to eight plastic, five-gallon containers in which to cool and ferment the mash

Getting Started. Recheck the interior of the drum to be sure that it is clean. It needn't be spotless, but it should be free of any oils, chemicals or other contaminants. Be sure the bolts on the clean-out door are tight. I find that it's easiest to first screw four bolts in each of the four corners of the door plate by hand, and then screw in additional bolts, evenly spaced. Once all are in place, I use a vise-grip to tighten them quickly, then the ⁹⁄₁₆-inch wrench to finish the job. Make sure the two-inch gate valve is also shut. Wrap the sheet of fiberglass around the drum, and secure it with wire or tape.

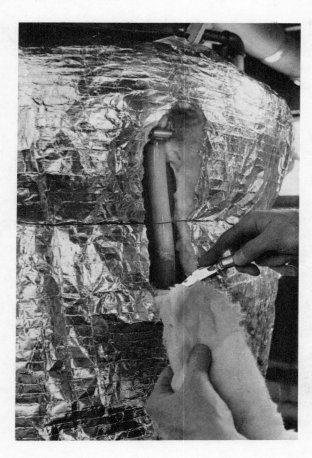

Use a fiberglass sheet to insulate drum during cooking. Secure the sheet with wire, then cut away a piece so sight gauge may be observed.

When secured by one bolt, the top hatch plate may be pivoted open easily.

There's a simple way to use the top metal hatch plate as a door. Face the drum and screw a single bolt into the corner of the plate nearest the front left top of the barrel. When stirring the mash, you can slide the plate open, using the bolt as a pivot; when not stirring, slide the plate back in place so that it covers the top hatch.

Pour 20 to 25 gallons of water into the drum, using either a hose connected to a faucet or plastic containers. With experience, you'll determine the exact amount of water you need. If you use too little, the mash will become excessively thick and difficult to stir and the likelihood of scorching will increase. If you use too much water, you'll dilute the mash unnecessarily. I'd suggest starting with about 25 gallons, and reducing this amount as you gain experience.

MASHING SAFETY CHECKLIST

- Keep a good, working fire extinguisher on hand at all times.
- Avoid overheating still with unnecessarily hot or fast-burning fires.
- Wear heavy-duty rubber gloves when handling mash or transferring it from still to plastic containers.
- Operate the still in a well-ventilated location.
- Ensure that the flue is installed according to fire safety codes.

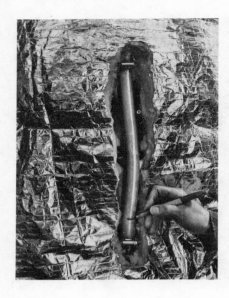

To save time, write a reference mark for the correct water level on your sight gauge.

Another tip you may find helpful: Once you have determined the amount of water that works best, notice this level in your sight gauge and mark the level with a dark, felt-tipped pen. For subsequent batches, you can simply refill the drum to that premarked level. Add the water to the drum first, followed by the corn. I use one bushel of ground corn (56 pounds) per 25 gallons of water.

To eliminate the need for repeatedly remeasuring 56 pounds, I fill a plastic container with 28 pounds, mark the level of corn, dump this amount into the drum, refill to the 28-pound level and then add this to the drum. In subsequent batches, I simply fill twice to the 28-pound mark and empty the contents into the drum. It helps to develop such shortcuts; this is a *labor*-intensive operation.

Lighting Your Fire. After adding the water and corn to the drum, add three teaspoons of *Canalfa*, mix well, light your fire and heat to 200° to 212°F. It will take about two hours to reach this temperature range.* Stir the mixture as often as possible. (Stirring is easier if you stand on a box or some kind of elevated support near the cooking drum.) Try to increase the heat of your fire steadily, to the point that you develop red-hot coals. As you reach about 190°F., ease off a bit; stabilize the fire so that once you attain about 200°F. it is under control. Hold at 200° to 212°F. for 15 minutes. Remove insulation.

---

*Cooking, distilling and cooling times will vary, depending on such factors as the temperature of the liquid at the start of these processes, the dryness and quality of wood used in the firebox, and the ambient temperatures at the still site.

Cooling the Mash. At this stage you're in a bit of a predicament: you've got about 30 gallons of slop, boiling along at more than 200 degrees Fahrenheit, and somehow you've got to reduce the mash temperature quickly to 150° to 170°F. I have not yet found a fully satisfactory way out of this fix. In winter, I shovel about half a wheelbarrowful of snow directly onto the hot coals. This smothers the fire reasonably well without too much steaming. In other seasons, I throw about five to ten gallons of water on the fire. This *does* cause steaming and sometimes it cracks the bricks I use to support the firewood.

With the flame retarded, I place a five- or seven-gallon plastic bucket beneath the two-inch gate valve at the base of the drum. I tilt the bucket slightly, so it's sure to catch the mash. Then I open the valve. Do this carefully, or the boiling hot mash will spew right over the bucket and onto the floor. Be sure you wear heavy-duty rubber gloves before you try this. Open the valve slowly. Even though this is a two-inch valve, you may have to prod the mash with a small stick before it starts flowing. Be careful. Once unplugged, the stuff will gush again. Although the valve plugs only temporarily, it happens often enough so that I don't believe any valve smaller than two inches would be big enough to drain the mash.

After draining the drum as much as you can through the gate valve, carefully remove the clean-out door and scrape out any additional mash from the bottom of the drum. You may find that much of the corn re-

One way to cool your fire is to simply shovel some snow onto the coals.

Place plastic barrel so it catches mash exiting from the two-inch gate valve.

This is what happens if I'm not careful when draining mash from the drum.

To pry the mash loose, you may need to insert a narrow stick into the gate valve.

Pivot clean-out door open then carefully remove mash from the bottom of the drum.

Distribute Canalfa, Gasolase *and yeasts evenly among the plastic buckets.*

mains on the bottom. I try to scrape approximately equal amounts of these solids into each of the six to eight plastic containers that I use to hold the mash.

Even after you remove the mash from the drum, it will probably require additional cooling to reach the 150° to 170°F. range. If necessary, use snow, ice water or ice packs; stirring also hastens cooling. Once 170°F. has been reached, add six teaspoons of *Canalfa*. (If the mash is in six separate plastic buckets, add one teaspoon of *Canalfa* to each bucket.) Hold for 30 minutes and continue to stir constantly.

After 30 minutes, cool the mash some more, down to 120°F., then add six teaspoons of *Gasolase*. (Again, divide the *Gasolase* as evenly as possible among the buckets containing your mash.)

Testing Sugar Content. Strain a small sample of the mash and pour the liquid into your hydrometer to establish the sugar content. Remember, read the figures on the Brix scale to determine sugar content; a sugar content of 10 to 15 percent is suitable for the start of fermentation.

Fermentation

There is no need to delay adding yeast to your buckets of mash. Simply continue cooling the mash down as quickly as possible to 100°F. and then 90°F. Once below 100°F., add 12 teaspoons of distiller's yeast. Mix well.

A simple fermenter with a plastic container, hose and milk jug with top re-moved.

Ferment the mash for three days or more. Try to maintain temperatures in the 85° to 90°F. range to achieve optimum fermentation. Stir every two or three hours if possible; if not, stir at least twice daily. When bubbling has ceased, the mash is ready for distillation.

For plastic containers to be suitable for fermentation, there must be a way for carbon dioxide gases to escape through the lid; at the same time, air must not be permitted to reenter the container. One solution is simply to fit a fermentation lock tightly into a hole punctured in the container lid.

Another solution is to insert a piece of plastic tubing into the hole and extend the free end of the tube into another container of water. The submerged free end will permit gases to escape but will not allow air to return to the mash container. Where the tube enters the plastic lid, I dab a water-flour paste to prevent air from entering the container.

Distillation

When I first started distilling with my 55-gallon drum, frankly, I was afraid the thing would blow up. I'd read stories and seen pictures of moonshine stills in shreds. I started my fire very slowly. I tightened down the top hatch, one bolt at a time. And I watched the temperature-pressure

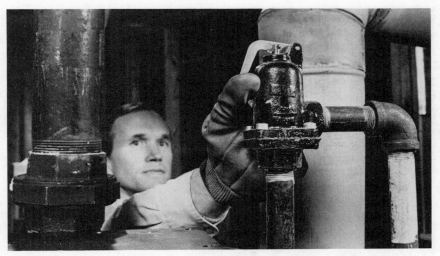

This type of pressure-release valve is checked by lifting its small arm.

gauge like a hawk. After only a few tries, though, I began to realize that my fears were unfounded. With a little care and diligence, the still can be operated safely.

The temperature-pressure gauge warns of excessive heat and pressure. As long as the temperature remains at or below 212°F., the pressure level will not go above zero. Another safety feature is the 30-pound, pressure-release valve. If for some reason the distillation column were to become plugged, excess pressure would be released through this valve. Check the valve before each run to ensure that it works. Simply lift the lever on the top of the valve to make sure it moves freely. Dangerous pressure should not develop *if solids are strained from the mash before the liquid is poured into the drum and if the drum is filled to no more than two-thirds capacity.*

DISTILLING SAFETY CHECKLIST

- Separate solids from mash prior to distillation.
- *Never* fill the 55-gallon cooking-distillation drum more than two-thirds full.
- Do not smoke. Alcohol fumes are flammable and potentially explosive.
- Be sure the column and copper condenser line are free of obstruction.
- Check temperature and pressure frequently.
- Make sure pressure-release valve works properly.

To separate solids, pour mash through a straining cloth pinned to the lip of a plastic container.

Getting Started. Once again, as a first step, pour the *liquid* from your mash buckets into the drum, then strain the remaining solids through a burlap bag or some other straining cloth. I find that a citrus fruit bag, cut so it forms a square piece of mesh cloth, works well. After pouring the solid-free liquid into the drum, check the sight gauge. Chances are, if you have prepared a one-bushel batch with about 25 gallons of water, the liquid will just barely show at the bottom of the gauge.

Next, fill your cooling tub or flake stand with water. Ensure that garden hoses are connected to the two fittings at the top and bottom of the drum. (Once the mash liquid reaches about 190°F., you will turn on the hose to circulate the cooling water. For now, though, simply fill the flake stand and let the water stand.) In the winter, circulation may not be necessary at all. Where I live in Vermont, it gets so cold that the flake stand water freezes. I have to use automobile antifreeze to keep my flake stand from turning into a drum of solid ice.

There's one final check to make before you light your fire: Climb up on a step-ladder and detach the ¾-inch copper tubing at the point (the union) where it connects to the 1-inch steel tubing. Blow through the copper line to ensure that it is free of any possible obstruction. Ask another

70

Before you start distilling, detach the straight end of the copper tubing and then blow through the tubing to ensure that it is not obstructed.

person to place a hand over the tail end of the copper coil. As you blow, your friend should feel air escaping from the coil.

Distilling. You are ready to light the fire in your still firebox. Again, try to generate a hot, even fire, with red coals as near to the top of the firebox as possible. With experience, you'll begin to learn the way your firebox operates. I almost never close either the door or the flue damper. Air circulation is needed to generate a hot fire.

I start distilling with the top hatch open so I can check the temperature inside the mash barrel and so I can stir the liquid. Stirring is not mandatory because there's almost no chance of scorching. But stirring does help to ensure that the liquid temperature increases evenly.

Once the liquid temperature has reached about 100°F., I screw all of the top hatch bolts down tight and watch the outside temperature-pressure gauge closely. What is needed is a temperature inside the drum warm enough to separate the alcohol from the alcohol-water liquid mash. A temperature of 190°F. to 200°F. on the temperature-pressure gauge near the drum top is about right to get the distillation going. It may take up to two hours to reach this temperature range.

As the temperature increases, vapor will begin to travel up through the distillation column. The thermometer near the top of the column will indicate increasingly warmer temperatures. Ideally, this temperature should remain in the 175°-to-180° F. range. After the first hour or so, temperature control will be difficult. Through experience, though, you can learn how to control your fire to achieve approximately the desired temperature ranges. Some stills include a cooling coil inside the top of the column, designed to keep the temperature down. That might be a good

71

Keep a careful record of temperatures and times when cooking and distilling. See Appendix D for a suggested checklist for the 55-gallon still.

A FEW DISTILLING TIPS

- Temperature control is very important and sometimes difficult to achieve. Cooking times needed for successful distillation will vary with the season.

- When the temperature-pressure gauge mounted on the outside of the 55-gallon drum indicates 160° to 170°F., begin to stabilize the fire. Don't add more firewood. Create a bed of steady, red-hot coals with little or no flame.

- Once the gauge indicates 190° to 200°F., remove the fiberglass insulation. Try to keep the upper column temperatures in the 175°-to-180° F. range. The column probably will overheat. One trick that seems to help is to wrap a garden hose around the upper column. Connect this same hose to the cooling flake stand. As you run cold water through the hose, it cools both the flake stand and column.

- Be patient. Even after you reach 190°F., it will be several minutes before distillation begins. Often, a tell-tale, brief, intermittent gurgling sound indicates that the first alcohol shots will soon emerge from the copper coil.

- The alcohol flows from the coil at a slow trickle. If the flow stops or becomes intermittent, add a bit more firewood to the firebox to keep the vapor moving through the still. Again, avoid overheating.

innovation. To keep the still design as simple as possible, I decided not to use such a coil.

Making Motor Fuel. As the temperature of the thermometer near the top of the drum approaches 190°F., you can expect to see alcohol trickling from the tail of the copper condenser coil within several minutes. As with the five-gallon still, I collect the alcohol in several plastic containers. I use a *proof* hydrometer to check the strength of the alcohol as each container fills. Usually, the first shots of alcohol will be 130° to 150° proof. After a while, the proof begins to decline.

Once the alcohol is consistently below 100 proof, I stop feeding wood to the fire and stop distilling. High-proof distillate (160 or more) is kept in plastic, one-gallon milk jugs; low-proof (100 to 150) is kept separately in other jugs. With practice, you may be able to distill about two gallons of 100-to-160-proof alcohol. Any alcohol below 160 proof should be saved, added to other batches with about the same proof and then redistilled. In about an hour or two, you can redistill low-proof alcohol to attain a fuel-grade alcohol suitable for modified combustion engines.

Still Maintenance and Fuel Storage. After completing either a cooking or distillation run, it's a good idea to clean your still promptly. I use a mix

Control the flow of water into the flake stand with the faucet at the base of the drum. Water exits through a hose near the lip of the drum.

of warm water and household detergent. Then I rinse thoroughly. Steel and heavy-duty fiber brushes make the job easier. Any waste mash is sent to the compost pile. Don't let the waste mash stand in the still overnight. It's unsanitary, and it's tougher to clean the still once the mash dries.

I store my alcohol in plastic or glass one-gallon jugs. I would not use plastic for extended periods, nor would I screw the container lids tight. Remember to keep a record of your alcohol production. The Bureau of Alcohol, Tobacco and Firearms requires it. The bureau rules also require that the alcohol be stored in a secure place.

CHAPTER 4

The ABC's of Pure Corn

BY JOSEPH EARL DABNEY

> Here's to Old Corn Likker,
> Whitens the teeth,
> Perfumes the breath,
> And makes childbirth a pleasure.
> —North Carolina folksaying

> There's gold in them there mountains,
> There's gold in them there hills;
> The natives there are getting it,
> By operating stills.
> —John Judge, Jr.
> *Noble Experiments, 1930*

It is a federal offense in America to distill alcohol without a federal permit or even to possess a workable, unregistered still.

But it wasn't always that way.

For 241 years of America's colonial and pioneer history, whiskey-making was an inalienable right of all citizens, an occupation free of federal restriction.* Which in part explains why it took on such major propor-

---

*Whiskey was untaxed in the U.S. up to 1791, and then from 1802 to 1862 (with the exception of three years following the War of 1812).

Reprinted by permission from *Mountain Spirits: A Chronicle of Corn Whiskey from King James' Ulster Plantation to America's Applachians and The Moonshine Life* by Joseph Earl Dabney (Lakemont, Georgia: Copple House Books, 1978). Dabney of Atlanta, Georgia has interviewed dozens of former moonshiners from across the South. Here, he describes the way these men worked in the remote, backwoods hills. Although the moonshiner's craft is often primitive and outmoded, this description has merit. Many who have tried to make alcohol fuel have borrowed the moonshiner's techniques. And just as moonshining is embedded in rural folk tradition, small-scale alcohol fuel-making has its origins in small-town rural life.

tions as a "cottage industry" on thousands of farms across the Appalachians and why, when whiskey-making without a license permanently became a federal offense in 1862, the result was widespread illicit moonshine activity, which has carried down to today.

Indeed, for the proud Scotch-Irish people who predominated on the frontier, whiskey-making inevitably was linked to freedom, and the various taxes imposed against it time and again in Great Britain and America were looked on as expressions of government tyranny. "Whiskey and freedom gang thegither," wrote the Scot poet Bobbie Burns.

To appreciate the full story of corn whiskey's history in America, one needs to know something about the nature of the enterprise, and the fundamentals of how alcohol is made. Since these fundamentals constitute a kind of practical guide to the making of corn whiskey, the consequences of violating the federal statutes in this regard cannot be emphasized too strongly. Violation of any one count of the federal liquor laws can result in a fine of up to $10,000 or a sentence of up to ten years in prison, or both.

Now, the manufacture of alcohol is a natural process, and man's intervention is required only to arrange the raw materials in the proper pro-

Moonshiners devised many ways to transport their illicit produce. Here, partially concealed barrels of alcohol are trucked through a city street.

portion and sequence and to apply the necessary heat and cooling along the way.

There are two basic steps involved: fermentation and distillation. Practically all foodstuffs, vegetables, fruit, and grain are fermentable—a natural phenomenon—and practically all of them at one time or another have been utilized to produce wine, beer, or alcohol. Beer, made from grain, and wine, made from berries and fruit, are the result of fermentation, and they end up with an alcoholic content—usually of sixteen percent or less.

During the fermentation process, the starches in the grain (or the fruit) are broken down through saccharification into sugars and then the sugars into alcohol. This process is speeded up greatly by the infusion of sugar, yeast and/or malt. Yeast speeds up fermentation by twelve hours. Sugar hurries it along additionally and increases the alcoholic yield considerably.

In whiskey-making, the basic fermenting mixture of grain, water, and other ingredients is called "mash." Mash containing a heavy proportion of sugar and yeast usually ferments in three to four days. However, pure corn meal alone, without sugar (though with the sprouted grain which is called "malt" and which is explained below), can take five to ten days to ferment and, if the temperature is not high enough, up to fourteen days or more.

Old-time distillers say the key to making a good run—a batch of whiskey—and drawing out the biggest quantity of alcohol, lies in the fermenting stage of the art, "mashing in" the grain, sugar (when used), and water, keeping it stirred, and applying adequate malt and/or yeast.

Soon after the meal, water, sugar, and malt are mixed in the mash box or barrel, carbonic gas bubbles begin rising to the surface, forming, along with some of the grain, a thick, foamy "cap." Some distillers add rye meal or wheat bran as an additional cap to contain the mash and speed fermentation. Later—the interval depending on whether sugar was used—the mash bubbles so much that it starts rolling, literally, and keeps on for a day and a half or two days. When the bubbles stop rising, and the cap disappears, the mash, now called "corn beer" or distiller's beer, becomes a soupy yellow and is ready to be distilled.

At this "high point," the beer contains ten or more percent of alcohol essence, and has a slightly sweet taste with a sharp sour tang. If it is not distilled promptly, it begins to lose its alcoholic content and turns more and more sour. Federal agents who come onto fermenting mash can tell immediately what stage of the cycle it is in, first by taste, and next by the "smack," the feel, of the mixture when the fingers are dipped into it. A clear liquid with little smack (little stickiness or tackiness) means it's about ready for the cooking pot and distillation. If the mash is a long way from

A 520-gallon "silver cloud" pot still captured in Cocke County, Tennessee. This still, fired by liquid petroleum gas, was said to resemble a silver cloud on a hillside, particularly at night under the light of a bright moon.

the distilling stage, agents usually elect to "save" the still and come back and raid it on the morning of the distillation, in order to capture the operators on the scene.

In earlier times, when sugar and yeast were not readily available, the distiller had to ferment his mash a lot longer than is common today. Some used honey or sorghum, known as "long sweetin'," which speeded things up a bit and helped the yield. In place of yeast, he made his own malt, sprouted grain which is dried and ground up. (Incidentally, it is a federal crime to grind sprouted corn or barley.) Malt, which was called "drake's tail" by pioneers, contains the enzyme diastase which converts the raw corn meal starches into sugar, through a saccharification process. Wild yeast spores that are present in the air multiply on contact with the meal in the mash and speed up the fermentation. In effect, the malt (and/or yeast) produce the souring catalyst for the mash in much the way that a pinch or two of sourdough will carry on the fermentation cycle for succeeding batches of dough.

To go from the fermented mash to alcohol itself requires the additional step of distillation. In this process the essence, or spirits, of the fermented liquid is separated from the water by being heated to the appropriate temperature. Wine heated to 173 degrees Fahrenheit gives off an alcoholic vapor which, when cooled and condensed back into liquid form, becomes brandy. Similarly, the distiller's beer produced by fermented grain-based mash produces grain alcohol — whiskey — when heated and vaporized at the same temperature. The higher the heat during the distillation, the greater the amount of water and impurities in the proportion of the final product.

The use of leftover soured slop from a still pot following a run — distributed back into the mash and sometimes referred to as "slopping back" or "dipping back" — continues the yeasting cycle for eight or ten subsequent runs, until the operator decides to "slop out" and begin anew. The first run of whiskey results in a sweet mash liquor. The subsequent slopping back distillations — usually with the addition of a lot of sugar and lesser amounts of grain — result in sour mash whiskey. "When whiskey gets good," says veteran distiller Hamper McBee of Tennessee, "is when it runs four times (on the same basic mash). Then you get some good whiskey — the kind you can drink without a carbide for a chaser."

Distilling being an essentially individual affair, there are many variations in the whiskey-making process. Here, in the words of different old-timers, are personal tips and suggestions on how they did it "in the old days."*

Putting in a Still

A retired whiskey-maker from Stillhouse Branch, North Carolina, explains the special attention that some corn liquor men paid to the water they used.

"The beer won't pay off as good if the water comes from a branch that's got touch-me-nots along its banks. . . . They denote hard water and hard

---

*The first comment comes from an interview by columnist John Parris of the *Ashville Citizen*. The remainder are taken from interviews [Dabney] conducted.

water won't make corn whiskey. For making moonshine, find yourself a branch where red horsemint grows. You can't go wrong. Another way to test the water . . . is to see if it will 'bead.'* All you do is take a jar of water from the branch and shake it up. If the bubbles rise when it's tilted, then you know you've got the kind of water it takes to make good whiskey."

Sprouting Malt

A seventy-three-year-old retired distiller from Laughingal, Georgia, tells of the value of using malt and how it is made. "We put our shelled corn in a tow sack and poured hot water over it and put it in a sawdust pile. Just covered it up. In about three days, it'd have stringers about two, three inches long. All tangled up. You just spread it out and let it dry in the sun for two, three days. Carry it to the mill and have it ground."

"In the wintertime," another maker from Rabun County, Georgia, said, "you have to dig a hole in the ground. Put your grain in it in a sack and pour warm water over it. Then cover it with straw and let it sprout. Sometimes takes four days. Then you put it out to dry and took it to an old water-powered mill somewhere and had it ground. Back when I was a boy, that was the only kind of mill they wuz."

Keeping Mash Warm and Using It

Keeping mash warm in winter was a problem. "We had a still one time on the quarry hill," a northeast Georgian told me, "and we buried twenty barrels. We got rye straw and hay and packed it around them. It was pretty severe winter, but the mash 'kicked over' in five or six days." (Many distillers found burying mash barrels also insured against excessive heating in summer. Too much heat will cause wild yeast "vinegar mother" to form, in effect killing the mash.) Another way to keep mash warm in winter is to bury the barrels or boxes in a sawdust pile. "The heat from that sawdust makes it work right along," an old-timer remembered. A third popular method was to bury vats in hot manure piles, "mother nature's heating system."

"When the mash starts workin', it rises up," declared an old-timer from Track Rock Gap, Georgia. "You gotta leave it eight inches from the top [of the barrel]. That malt'll come up there thick. If it's still workin', it's 'wild in the still.' That kind 'pukes' [meaning the beer belches through the vapor line prematurely before it is distilled]. Got to let it settle down. It's

---

*See the glossary for terms that need further clarification.

Mash barrels lined up for a North Carolina still. The distilling pot is the stave-type barrel, followed by thump keg and flake stand. Note tent-like camouflage at rear.

all right when it begins to spot [clear] just a little. You can see the clear spots. Then when it settles down . . . down . . . down . . . the cap's about dropped on it. It's ready to run likker! Ready to run likker! You've got eight hours to cook it. Otherwise it'll go bad on you."

Firing the Still "Just Right"

Firing a still furnace is an art in itself, because the idea is to keep the heat close to the alcohol vaporizing level — 173 degrees (F.) — but not much higher. An old-timer from near Blairsville, Georgia, recalled this stage:

"I've seen faar at night just lap its tongue out in front of the still. The blaze crosses itself when you're faarin' one hard. You get that furnace throat just exactly right, it'll cross its tongue.

"After a while — it don't take too long — your furnace is good and hot, you'll see a little steam look like comin' out of the worm. Next thing you'll see directly a drop, maybe two or three drops. Then it'll piss just a little. After a while, when you begin to see drops comin' fast, you say, 'She's gonna get riiiiiiiiiggghhhhhhhhhttt now.' Directly she'll start pourin' out of the worm. *Then you're makin' likker.* You don't want to faar it too hard. You can puke a doublin' if you do. You want to run it slow and steady. That whiskey'll get out there [out of the worm] and hold its tail, like a saddle horse kinda. It'll get out there just as steady. . . ."

Moving From a Single Run to a Doubling

After the distiller has run eight barrels of mash through the still to make singlings, he is ready to start on his doubling run. The same distiller from Blairsville, Georgia, quoted above, describes this phase:

"You run them singlin's 'til as long as they got any strength in 'em, then you put your hands under the worm, rub 'em and inhale. When they smell right sour, the alkihol's out. It's ready to change boils, and you refill the pot with those singlin's for a doublin' run.

"But first, you draw your fire and clean your still. Ever' little spot. Everything's got to be clean about likker. Wash your still . . . wash your vessels after a run. If you're thumpin', wash that thump keg every time you run a boil of likker. *Wash* that thing. *Clean* that thump barrel. *Wash* the heater box. *Wash* 'em connections. That don't leave no feinty taste. When that bead breaks when you run them low wines in there, it gets down to a funky scent, leaves a bad odor in the connections, and on your next run, that likker's gonna knock that bad odor out and that's in your likker there, you see."

Telling When Whiskey is About to Break at the Worm

The maker keeps a close watch on the distillate coming out of the worm to make sure he pulls away his tub when the whiskey breaks at the worm, when the proof drops sharply. A one-time maker from Habersham County, Georgia, had a sharp eye for the break. "I could stand as fer as from here to that door yander and I could tell just when it would break. It quits runnin' nearly . . . the power and the strength stops—drops to a stream a little bigger than a match stem. I could take my finger and wet it that way and put it to my mouth and I could tell just the second I tasted it . . . I didn't have the power."

Another maker from Union County, Georgia, remarked:

"When thump likker shows its low bead, it ain't long before it's through. Lots of times I could tell when the bead broke because it changed that twist. I'd say, 'It's broke boys,' and go over and test it and it'd be dead."

Proofing Whiskey By Its Bead

Next comes proofing the whiskey by its bead, as explained by a former moonshine hauler—a tripper.

"If it's high proof—say 115 to 120—a big bead will jump up there on top [of the whiskey] when you shake it. If the proof is lower, the bead goes away faster and is smaller. Hand a mountain man a pint of whiskey and the first thing he'll do is shake it. The longer those beads stay on there the higher the proof.

"There's a way of putting a false bead by using beading oil or lye. You put the oil in when the likker first comes out of the still. The way you tell . . . when the bead hops all the way out and sits on top of the surface, it's false. A true bead will stop half in the likker and half out on top."

The proof of whiskey today is estimated this way: Its advertised proof figure is twice the amount of the alcohol content. If it is labeled 100 proof, this means it is 50 percent alcohol. The British centuries ago established proof at 57.1 percent alcohol. It had been found that whiskey with that proportion of alcohol, mixed with gunpowder, gave off a steady blue flame. An ancient British dictionary described this as "gunpowder proof,"* which the British government adopted in 1816 as being 100 proof. The U.S. copied the principle but, to make it easier to calculate, ruled 100 proof whiskey to be 50 percent alcohol. The British still have proof of 57.1 percent alcohol by volume, which translates to 114.2 proof American!

Some distillers who are not confident of their estimates of the bead toward the end of a run have been known to throw a cup full of whiskey into the furnace. If it blazes up well, they are assured the proof is still strong. Early revenue officers of the American frontier used something called a Dicus hydrometer. In the case of most distillers in the Appalachians up through the early 1900s, proof was checked with a proof vial—a small bottle-like device, to which they would add drops of water to the whiskey. "You'd get in two-thirds full of likker," a retired maker remembered, "and put twenty drops of water in it. If the twenty drops

---

*Early American Whiskey historian Harrison Hall of Philadelphia described "gunpowder proof" this way: "Pour a small quantity of spirit on a small heap of gunpowder and kindle it. The spirit burns quietly on the surface of the powder until it is all consumed, and the last portion fires the powder if the spirit was pure; but if watery, the powder becomes too damp and will not explode."

killed it, knocked the head off of it, that meant it was 100 proof likker." In general, though, most distillers merely shook a vial and judged the proof by the bead.

Straining Fresh Whiskey

Filtering whiskey—purging it of its impurities—also has a technology of its own. Although most modern-day moonshiners shun filters, many makers of earlier days used "bootleg bonnets," to strain the fresh whiskey. They would tack the hats across the top of the keg under the worm. But the most popular method of the old-timers was to use homemade charcoals. Here's how Peg Fields, a retired distiller from Pickens County, Georgia, did it:

"We had a big funnel set in a two-gallon bucket, and we placed it right under the worm. We faared with hickory wood most of the time. We'd burn them faar coals and wash 'em off in the branch. We'd put a flanigan cloth or two in the funnel bottom and fill it with them coals, the whiskey coming from the condenser down through the coals into the bucket. If you used barley malt [to make the liquor], the whiskey left grease on the coals. Wasn't much left when we used corn malt. But those coals cleaned the whiskey up and left it just as clear as could be."

Distillers on the Cumberland Plateau in Tennessee, who specialized in improving raw whiskey by "rectifying" it (another term for purging), set up elaborate filter systems. They would take an empty keg and place it in a layer of felt, a layer of maple charcoal, a layer of sand and a layer of gravel. Then they primed it with ten gallons of high proof whiskey, sending the whiskey through the several layers. But most illicit distillers never went to such elaborate extents. (Some distillers even purged impurities by placing a charred peach seed or a few spoons of ground charcoal into a jar of whiskey.)

Aging Corn Whiskey

Most corn whiskey over the years has been sold fresh and unaged—seldom over a week old—which meant that it was clear, "white lightning," as it was called. But many of the registered, family-operated "government distilleries" that proliferated following the Civil War, charred their barrels in the bourbon tradition to age their whiskey, giving it color and a smoother taste. A retired tripper from Cobb County, Georgia, remembered how his grandfather did it in the late 1800s in Dawson County:

"He burned the barrels out with still alcohol. He'd take high shots and slush 'em around in the barrel. Set it on fire and let that blue flame shoot out. Then wash it out with boiling water and clean out the ashes. It would char it down about a half inch deep. You always charter [sic] old barrels that way to burn out the old fusel oil."

But makers virtually always aged their "drinking likker." Different people had different ways. From a retired maker in Jasper, Georgia:

"You take a white oak or a hickory tree and split off little pieces about half as big as your finger. Take six or eight pieces and fry 'em good and brown in an old pan outside. Drop five or six in a gallon jug of whiskey. Ever time you go by it, just shake it a little. First thing you know, it's as pretty and red as any bonded whiskey you ever saw."

From a distiller near Greeneville, Tennessee:

"You put your whiskey in a charred keg. Go to the keg about twice a day and shake her real good. In three weeks, it'll be cherry red. That's some of the *best* likker that you've ever stuck in your mouth. She'll char more that way in three weeks than hit will settin' right there perfectly still in six months. That kerreck."

A copper pot still: vapor flows from the cap arm into the thump keg, then into the heater box at rear, and from there into a copper coil condenser. In foreground are 220-gallon mash barrels.

The Fine Art: Making a Run
of Pure Corn Whiskey

Toward the end of my research into corn whiskey lore, I was fortunate to meet and interview Arthur Young, a seventy-year-old native of the Smoky Mountains. During his youth, Mr. Young learned all about the pure corn art which was done, of course, without the use of any manufactured ingredients — that is, yeast, sugar or store-bought malt. The only ingredients used in the old days were corn or rye meal and home-sprouted and ground malt, and, of course, good soft water from a clear mountain stream. Mr. Young is also a musician, and before he turned twenty quit making liquor and devoted his energies to, among other things, fiddling on a Stradivarius violin (made in 1700) that he inherited from his father and his "greater grandfather" who brought it over from England. In previous visits to the Appalachian foothills of far northeast Georgia, I had heard about Mr. Young's intimate knowledge of the distilling art (as well as his musicianship), but it took some doing to find him. He resides in a little mountain farming community called Tate City, deep into the Appalachian ridges along the headwaters of the Tallulah River. When I reached his home, situated beside a one-lane dirt road on a bluff overlooking the Tallulah River valley miles from a paved road, I found Mr. Young sitting out in his front yard.

"I was born in 1903 in the Smokies under Clingman's Dome, North Carolina," he told me. "I fooled with whiskey from the age of sixteen until I was nineteen. Quit fooling with it when I got another job. Drove horses for thirty-five years for these old logging companies. I quit whiskey. Wasn't nothing in it, you couldn't get nothin' fer it anyhow. Likker then sold for $1.50 a gallon. Likker that would sell today for $24.00 a gallon. Course most likker today is chemicals. It's never seen no still. That's the kind of inspectors the government's got on it now. Back in my day, they wan't any government likker. Whole United States was dry. It was all blockade likker."

Mr. Young explained how he and his associates made corn whiskey in the old way, singling and doubling with a copper pot still. First of all, he had his copper pots made by a skilled still-maker. Then he selected a secluded spot beside a "bold" stream of water and built up a furnace from rocks gathered from the water bed. He fashioned his condenser on the spot, making it into a worm.

"First we got a copper tube three-quarters or an inch in diameter, about sixteen, eighteen feet long — long enough so that when it is curled, it would go from the bottom of a fifty-gallon barrel to the top. We poured

Here's a simple, two-stage pot still fabricated by a blacksmith from central Illinois.

either sand or sawdust in the tube to keep it from crimping and twisted it around a stump. Placed this worm inside the barrel and fitted the top of the tube to the cap arm from the still. We never did use a thump keg. Went directly from the pot to the worm. We put a trough into the barrel — the flake stand — and ran cool water through it all the time."

Here follows Mr. Young's description of how he went about the process of making singling and doubling (or double and twisted) whiskey on a fifty-gallon pot still:

"If a man had a fifty-gallon still, he'd need about eight bushels of meal — ground from choice white corn only.

"In addition to your still and worm [and flake stand barrel], you need eight fifty-gallon barrels. To start off with, you put a half bushel of meal in each barrel. Then you 'cook in' the other half of the meal — four bushels — in your still pot. Just heat it up and make a mush of it, like cornmeal batter. Then you divide that back into the eight barrels. So what you'd have in each of those barrels would be a half bushel of cooked meal and a half bushel of raw meal.

"You leave it a couple of days, then you have to go and break it up, and thin it with water, and mix it up with a stir stick. At that point, you add a couple of gallons of ground corn malt and a gallon of rye meal. Some folks sprinkled rye meal on top to form a cap and keep the mash warm. Others just stirred it up. It'd come to the top soon anyway — all grain does. Six

87

days later, when that cap falls and the top gets clear, the mash had become 'beer' and it was ready to run off."

One by one, Mr. Young said, he distilled all eight barrels. From each of these singling runs, he said, you could expect to get six to eight gallons of "singlings," or low wines. Then, "you take those fifty gallons of singlings and cook them through the still again. That's your doublin' run; that's when the alkihol comes. It'd be real *alkihol*. Be high proof, too high to bead. That's grain alkihol. Those first shots be 150, 160 proof. As it continues to distill, the whiskey proof gets weaker and atter while it comes down to a bead [about 120 proof].

"Finally, it comes to a good bead, about 100 proof. Corn whiskey has a lot smaller bead than this sugar whiskey nowadays. Sugar puts a big coarse bead on it. Now pure corn whiskey comes down to a *fine* bead. About the size of a number six or number seven shot, not much bigger than a bb shot. Sometime later, the whiskey breaks at the worm, as they say. It starts smelling and tasting sour. That's below 90 proof. You're supposed to take all the whiskey that comes out from then on — the backin's — and put it in a barrel with the singlin's for the next doublin' run."

From fifty gallons of singlings, the distiller expected to get between sixteen and twenty gallons of "doubling likker." "I just always figured it'd make two gallons and a half to the bushel of meal," Mr. Young said. "Then you'd 'fourth' that with water, which in this case [eight bushels at two and a half gallons meant twenty gallons of whiskey] would amount to five gallons of water, bringing your yield to twenty-five gallons of doubling whiskey. Some people cut the proof down with backin's," Mr. Young said, "but we didn't. We took cold spring water and poured in that. Cuttin' with water makes the best likker . . . the sweetest-tasting likker. Lots of folks would boil the water first and let it cool off, then added it to the whiskey."

While whiskey-making, illicit style, essentially was a case of "helping nature along" — and seemingly a joy ride — it was also a tough, backbreaking occupation, requiring a lot of plain hard work and steel nerves. Many older mountain men with crooked backs can attest to their younger days of "rawhiding" sugar and meal into tough mountain terrain and then barrels of whiskey on the opposite trip out.

As Mr. Young observed: "I usta hear people complain that these whiskey men ought to get out and work for a living. The fact was, *they* were the ones were *really* working already."

I remember asking an old-timer from Mountain City, Georgia, if he enjoyed his former life as a moonshiner:

"Oh, it was excitin' at times, but it's not a life of sunshine, I'll clue ye. *It's nothing but hard work and the very hardest.* You stay out there and take all kinds of exposure. If you ain't got a shed over you and if it comes up a rain, you stand right there and take it. If it snows, you stand right there and take it. Lot of time, maybe you got no way to ride and you too fer back in the mountains to walk out and you just got to stay and take it."

Hamper McBee put his share of hard labor making moonshine in the Cumberland Mountains northwest of Chattanooga. On a record he cut, called *The Cumberland Moonshiner*, Hamper commented about the image of the lazy whiskey-maker:

"Ain't no lazy man gonna make no whiskey . . . I hear a lot of women at our home say 'so-and-so too lazy to work, all they do is make whiskey.' They ought to try it sometime if they think it's a snap. Ride the sugar down the mountains and it wet, slippery, and you're falling and a stumbling and getting down there a chopping wood and it wet and trying to get a fire and run that stuff and don't know whether you're agonna run into the revenue or not and then have to pack that stuff back out of them mountains. It's a blooming job. . . ."

Against these odds, extremely hard work, the ever-present danger of

Many materials are used for homemade stills. This old truck radiator served as a condensing unit. Often leads or impurities in these condensers caused serious, and even fatal, poisoning.

A black moonshiner at work at his still in Southern Pines, North Carolina. The still is a metal pot with a thump keg and condenser.

being caught by the law and the relatively low compensation — why have men across the South continued to make whiskey illicitly over the years?

A distiller from Bartow County, Georgia, offered one rationale:

"Moonshining gets into your blood just like sawmilling and digging gold. Once a gold digger, always a gold digger. Once a blockader, always a blockader at heart."

Appalachian researcher John Gordon, writing in the *Georgia Review*, had another explanation:

"Mountain people are action seekers. They live episodically and they live for adventure. Moonshining is for some of them the ultimate adventure."

CHAPTER 5

The Zeithamer Still,
Alexandria, Minnesota

By Oren Long

ALEXANDRIA, MINNESOTA. The idea that farmers can produce their own
alcohol fuel is no longer novel. On-farm production of alcohol fuel is past
the talking stage. Industry is producing farm-sized alcohol plants, and
farmers are buying.

Two who appear to be making progress in putting theory into practice
are Archie and Alan Zeithamer, father and son dairymen at Alexandria.
Not only have the Zeithamers built a fully licensed, alcohol-fuel distillery,
they have produced similar plants for sale.

Zeithamer describes his plant an "in-and-out" type of operation. By
this, he means he can do chores and other nearby farmwork while producing alcohol. With continuous operation, the maximum plant capacity
is about 50,000 gallons a year.

Zeithamer adds, though, that he thinks most farmers will produce
alcohol only in the winter months, or during harvest time, using high-
moisture grain as their basic raw product for alcohol production.

When it comes to producing alcohol, some farmers have a built-in ad-
vantage; they have relatively inexpensive supplies of readily available
fuels to power their stills. These include wood, waste oil, old tires, sawdust, and crop residues. Even manure can be used to produce methane gas
with a digester.

Zeithamer estimates that it takes about 1 cord of wood (17 million Btu)
to produce one batch of alcohol, or about 400 gallons. One cord is about
equivalent, in Btu, to the residue from ½ acre of corn stalks, 1 acre of

---

This chapter, adapted from an article that appeared August 4, 1979 in *Kansas Farmer*
magazine, is printed by permission of the author.

stubble from average yield milo or 1½ acres of straw from average yield wheat.

Most farm stills require some form of supplemental heat to ensure even, specific temperatures during the cooking of corn mash and distillation of alcohol. Conventional fuels such as gas and oil will probably be needed, in addition to fuels such as wood and crop residues.

A Promising Future for Alcohol?

Many believe alcohol fuel has a promising future as a supplementary, alternative source of energy. At least one national magazine believes there may be a widespread grassroots movement developing in the schools: "As soon as the word on alcohol filters down to the hot car crowd, every high school auto shop in the country will build its own moonshine still."

And one knowledgeable expert told me, "I believe that in five to ten years, new enzyme research will make possible the development of small, inexpensive alcohol kits, powered by solar collectors, which will enable many urban and suburban dwellers to produce enough alcohol from grass, leaves, garbage and garden refuse, to power their lawnmowers and motorbikes and perhaps even a small car to and from work."

One side benefit of alcohol production is that the waste or spent grains can be used as a high-protein feed. The residues from making alcohol contain almost all the crop nutrients, and more of the protein, found in the original raw agricultural product. In my opinion, such residues could be added directly to the soil with no ill effects on soil or crop quality. A short stay in the compost heap would quickly break down any residual alcohols into carbon dioxide and water.

The Zeithamer Still

Heat sources for the still are two fireboxes, one under the 4,000-gallon fermentation-cooking vat, the other under the boiler and distillation column. The flue for both is a concrete block chimney that exits directly through the center of the barn roof.

The fireboxes have steel lids that can be opened to add wood or check the fire. The firebox under the fermentation-cooking vat has concrete block sides that extend halfway up the vat. The firebox under the boiler is smaller with a steel door for adding firewood.

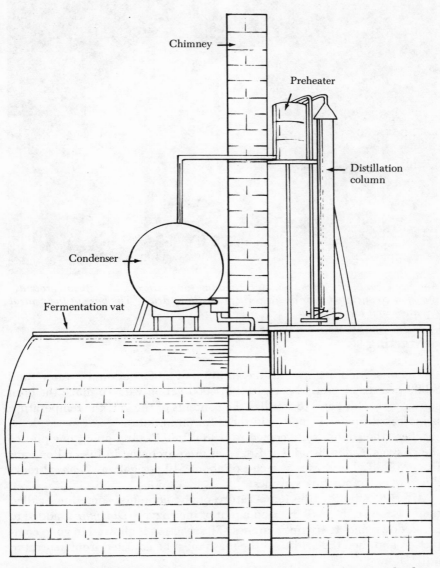

The Zeithamer still. There are two fireboxes at the base, one heats mash, the other the boiler and distillation column; after straining through a screen, the mash liquid ("beer") is pumped from a below-ground beer vat up to the pre-heater. From there, the beer passes into the distillation column; and, after distillation, the alcohol vapors travel on to the condenser. Alcohol is stored in separate containers.

Another view of the Zeithamer still, showing the firebox in the left foreground, the square concrete boiler and a vertical auxiliary drum. The base of the central chimney is at top of photo.

Starting

The operation begins with the addition of 2,000 gallons of water* and 7,000 pounds of ground corn to the 4,000-gallon vat. To make the corn starch more fluid and to sterilize the contents of the vat, the temperature is increased to 200°F. and held there for 30 minutes.

Then, the contents are cooled to below 190°F., and about one pint of the enzyme *Tenase* is added. Enzymes are purchased in five-gallon cans from Miles Laboratories at a cost of about $10 per gallon. *Tenase* breaks down one- to four-link starches.

After 90 minutes, the temperature of the contents, sometimes called mash, is lowered to 140°F., and a pint of the enzyme *Diazyme* is added to break down one- to six-link starches. The reaction is allowed to go to completion and then the acidity of the mash must be adjusted from neutral to a pH of 4.

Sulfuric acid or hydrochloric acid is added to adjust the pH of the mash, and the pH is determined with specially treated papers. When placed in the mash, the papers give off a color, indicating the pH. Lime

---

*Archie Zeithamer recommends that those interested in farm-sized distillation units have a careful, professional water analysis made. If unfavorable water conditions exist, consult a fermentation expert or enzyme manufacturer for assistance.

can be added if the pH needs to be adjusted upward. A hydrometer is used to test the sugar content, or alcohol potential. Finally, an iodine test should indicate that most of the starch has converted to sugar.

Fermentation

Yeast, the agent that feeds on sugar to produce alcohol, is added to the mash at a rate of about four pounds per batch. The yeast is premixed with water to hasten the start of fermentation. (The manufacturer, Anheuser-Busch, may recommend different dosage rates for the yeast.)

For optimum yeast activity, the mash temperature is reduced to 86° from 90°F. The yeasts used by the Zeithamers can tolerate about 11 percent alcohol by volume. The fermentation process gives off heat, which helps keep the building warm.

The progress of fermentation, or the conversion of sugar to alcohol, is monitored by taking readings from a saccharometer and from a hydrometer. A manhole on the vat permits access to the mash. (The hole is also used for filling the vat.) The lid fits loosely to allow carbon dioxide gas, a

Mash cooking and fermentation occurs in this horizontal, 4,000-gallon vat. Note the condenser mounted on top and the electrically driven stirring device and chain at the end of the vat.

fermentation by-product, to escape. Fermentation is complete when the specific gravity of the mash is close to 1. It takes about 36 hours. Temperatures are monitored with two meat thermometers mounted in the end plate of the vat.

Zeithamer says there is no need for stirring the mash during fermentation. An electrically driven stirring device is used during mash cooking. An electric motor drives a steel shaft mounted on the center line of the cylindrical fermentation vat. The shaft has 14 stirring paddles, and it turns at a rate of 12 revolutions per minute.

Separating Grains and "Beer"

The mash exits from the vat via two 2-inch drain pipes. The pipes convey the mash to a trench covered by a screen; the screen separates the mash solids or grains from the liquid, often called *still beer*. The wet grain solids go directly to cattle as a high-protein feed. Recovered from the vat, the grains have a protein content of about 38 percent.

The beer vat is a second 4,000-gallon, used gasoline storage tank buried in the concrete floor near the fermentation vat. It receives the beer from the concrete trench under the screen. The beer contains about 7 percent alcohol by volume.

Preheating

One of three ordinary water pumps used in the still elevates the beer to a 55-gallon preheat tank mounted near the top of the six-foot high distillation column. Collecting some heat from the pipe that carries alcohol vapors from the top of the column to the condenser, the tank preheats the beer to about 110°F. A valve regulates the beer flow to the column through four nozzles. A manifold mounted halfway up the column directs the beer to all four nozzles.

Distillation Column

The 12-inch diameter, steel distillation column has 40 perforated plates that are mounted horizontally. The 12-gauge steel plates are punched full of $3/16$-inch holes located, from center to center, $5/16$ inch apart. The hole punching cost about $305 at a steel fabrication shop.

The plates are fitted loosely inside the column and are separated by

The still chimney passes up through the center of the Zeithamer barn. To the right of the chimney are the pre-heater and distillation column.

two-inch metal spacer rings. (From two- to four-inch spacing can be used.) The beer enters halfway up the column; 20 sieve plates below this point are called *strippers*, and the 20 plates above the midpoint are called *rectifiers*. Temperatures are monitored with three cooking thermometers mounted along the length of the column.

Condenser

Alcohol vapor with a temperature of about 175°F. leaves the top of the distillation column through a tube. The tube passes through the beer pre-heater and continues to the condenser. The condenser consists of two heavy-duty, tractor radiator cores, mounted inside a large 1,000-gallon cooling tank. The alcohol vapor enters the top core and cools or condenses into liquid alcohol as it continues down through the other. Zeithamer says the cores are difficult to drain, and that a copper coil would be preferable. The alcohol flow may be viewed through a window.

Sampling Alcohol

A sample valve is a device fabricated so that each time a tap is turned a counter unit records the movement. (Sometimes these devices are required by the Bureau of Alcohol, Tobacco and Firearms.) Movement of the tap allows 10 ounces of alcohol to pass into a chamber where it may be tested for proof.

Proof is established with a second hydrometer called, appropriately enough, a *proof hydrometer*. When below 160 proof, the alcohol is unsuited for fuel and it is diverted into a reflux tank. The alcohol is stored temporarily in this 50-gallon tank, then recycled.

During the production run, low-proof alcohol is fed into the top of the distillation column for redistillation. A separate manifold distributes the alcohol to four nozzles. When the still is operating at optimum efficiency, it produces 185-proof alcohol, a grade of fuel that can be used in many cars with minor modifications.

Alcohol Storage

The alcohol storage tank has a capacity of 1,000 gallons. The tank has a glass tube sight gauge and a locked cap, to prevent alcohol removal until a BATF agent opens the lock and adds a denaturant. Denaturants, such as

Alcohol produced by the Zeithamer still is stored in a 1,000-gallon tank. The Zeithamers use denaturants to make the alcohol unfit for drinking.

wood alcohol or methanol, make the alcohol unsuitable for drinking. Records are kept in a nearby desk and cupboard.

The alcohol storage tank has air vents, located higher than any other part of the distillation equipment and designed to prevent illicit alcohol removal.

Boiler and Slop Tank

The boiler is nothing more than a tank of water between the bottom of the distillation column and the firebox. As water separates from the beer, it dribbles into the boiler and excess water flows into the slop tank. There is no control over the amount of steam produced, but there is little pressure in the boiler.

The slop is pumped from the tank back into the fermentation vat to be used as part of the next production run. Any tank, such as a used 4,000-gallon service station tank, would be suitable for a still slop tank.

CHAPTER 6

The SWAFCA Still,
Selma, Alabama

SELMA, ALABAMA. This city on the red-clay banks of the Alabama River, this historic Confederate munitions depot, this starting point for Martin Luther King's march to Montgomery, this jarring mix of southern majesty and abject poverty seems an unlikely spot for making alcohol fuel. But the mixture was right for Albert Turner and the South West Alabama Farmers' Cooperative Association (SWAFCA).

In 1978, this 2000-member, black farmers' co-op started making alcohol fuel from corn. Turner, a southern civil rights leader, was looking for ways to revitalize the cooperative and to aid its financially strapped members. Many had been left to fend for themselves after they lost plantation jobs during the civil rights turmoil of the 1960s.

There were other reasons the decision to make alcohol fuel seemed right. Gasoline prices were escalating; and the co-op had been faced with three, successive, poor growing seasons, leaving a shortage of revenue-producing crops. In Turner's mind, though, the nearly diasterous crop failures carried a silver lining—hundreds of pounds of waste crops, unsuited for the market place but ideal as raw products for making alcohol fuel. Under these conditions, he and a small group of co-op members built an inexpensive fuel plant.

The SWAFCA Plant

The SWAFCA plant stands next to the south end of the co-op's large, steel-roofed warehouse, about ten miles west of the city of Selma, just off Interstate 80. The plant and a one-story co-op office building are in a scrubby field, edged by pine groves so typical of this part of the country. The plant has these six basic parts: three large, box-shaped combination

101

The SWAFCA plant has a tall, central distillation column and three large, rectangular cooking-fermentation tanks.

cooking-fermentation tanks; a steam furnace; a 10-foot distillation column; a water-cooled condenser; storage tanks; and a secondary system for recycling low-proof alcohol. The plant was built with new and used materials that cost about $6,000; Turner said he paid a man $7,000 to weld the parts together.

The SWAFCA plant represents the culmination of efforts by some early pioneers in the field of homemade alcohol fuel. Although much of the technology has been improved upon, the plant suggests what a small group of committed citizens can accomplish.

Basically, this plant is an adaptation of designs and operational techniques used by both moonshiners and industrial distillers. The SWAFCA plant works about the same way as a design known to moonshiners as a "North Georgia-type still." Steam generated by the furnace is piped into the combination tanks where it first precooks a corn mash. After cooking, the mash remains in place and the tanks become fermentation vats. Then, after fermentation, the mash is carefully reheated. At 173°F. pure alcohol begins to separate from water and to vaporize. The vapors travel upward into the distillation column where they're further "stripped" of water and impurities. Next, the vapors travel into the condenser, mounted horizontally at the top of the column. Here, the vapors re-form as liquid. The cycle is complete when the liquid alcohol flows by gravity to storage tanks below.

One feature of this plant is that there is no requirement to move the mash to different tanks during the cooking, fermentation and distillation stages. All steps are accomplished in the same combination tanks. And there's almost no requirement for agitation or stirring equipment. Steam does the job. When more agitation is needed, Turner slips a high-pressure air hose into the mash. The secondary distilling system (essentially, the major parts of a smaller still) recycles low-proof alcohols. Alcohol samples are taken as distillation progresses in the main plant. Simply by adjusting a few valves, low-proof alcohol can be returned to a second distillation column and then the main condenser. The entire plant is "gravity fed"; there are no pumps or electrically driven parts. Turner said he can produce 185-proof alcohol on the first run. And, he added, alcohol with a proof as low as 160 works—poured straight into his 1975 Ford pickup. When in operation, the still usually produces 160- to 170-proof alcohol.

The Plant Design

Turner—a busy man, in the midst of answering phone calls, talking to farmers, operating a fork lift and inspecting cucumbers at the warehouse—took the time to describe the plant in more detail. First, he talked about the basic materials and parts of the plant, and then the steps he follows when cooking, fermenting and distilling corn mash.

The Steam Furnace and Combination Tanks. As mentioned earlier, mash processing occurs in three tanks, heated by steam. The tanks and steam furnace are arranged in a row at the south edge of the warehouse. The warehouse walls are open at this end, allowing plant operations to be viewed from inside the warehouse, protected from inclement weather. Because the tanks are in the open, however, it's sometimes difficult to maintain optimum fermentation temperatures during cold weather. Most alcohol plants of this size are enclosed.

The box-shaped tanks and furnace are made of $\frac{3}{16}$-inch-thick sheet metal. Each 5,000-gallon tank is 6 feet high, 18 feet long and 6 feet wide; the 1,500-gallon steam furnace measures $6 \times 8 \times 6$ feet respectively. These and other dimensions may be scaled up or down so the plant meets any capacity requirements, Turner said. In addition, to decrease capacity, the SWAFCA plant may be operated using only one of the three combination tanks. If you have detailed questions about metal dimensions, thicknesses and strengths, it would be advisable to contact a consulting engineer. (See Appendix C for a list of distilling plants and engineers.)

The tops of the tanks and the furnace are pyramid shaped; in the case of the tanks, this shape channels alcohol vapors to ducts leading to the dis-

The still furnace is mounted on I-beams and insulated on three sides with bricks. Steam travels from the furnace, by pipe, to adjacent cooking-fermentation tanks.

tillation column. The top of the furnace tapers to a stovepipe. The furnace is mounted on I-shaped steel beams and is insulated on three sides by brick walls. The beams keep the furnace above its heat source, a wood fire; the walls help retain heat within and around the furnace. All four sides of the furnace are reinforced with a band of angle iron and the three tanks are set in concrete.

Water reaches the steam furnace via a heavy rubber hose connected to a 1½-inch metal pipe at the base of the furnace. Water is controlled by an iron body gate valve at this connection. A plastic-tube sight gauge mounted on the side of the furnace permits monitoring of water levels. Steam exits through a 3-inch pipe that travels down from the top of the furnace to a point near its base. At that point, an elbow joint turns the direction of the pipe 90 degrees, toward the row of three combination tanks.

Steam travels along the 3-inch pipe to each of the three tanks. Smaller, 2-inch pipes connect the main, 5-inch line to the tanks. If reduced tank capacity is desired, steam to any one of the three tanks can be shut off by simply turning an iron body valve on the appropriate 3-inch line. Once inside, the piping is reduced further. The 3-inch pipe is welded to U-shaped ¾-inch piping, which distributes the steam throughout the tank. The steam exits through pipe "legs" that point straight down to the bottom of the tank.

At the center of the top of each tank, there is a square hatch made of

104

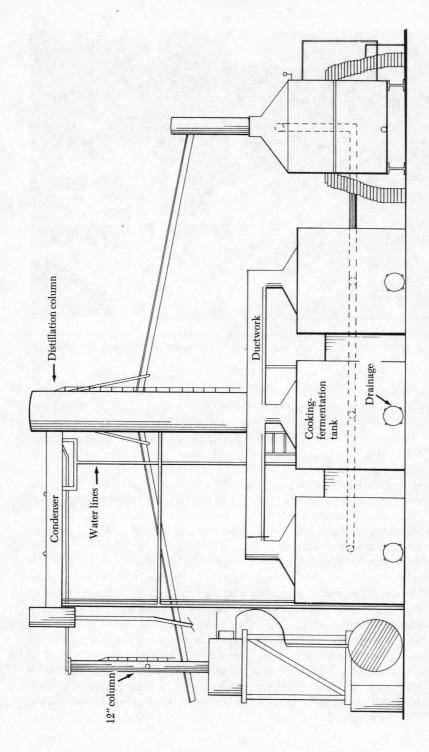

Condenser

Water lines

Distillation column

12" column

Ductwork

Cooking-
fermentation
tank

Drainage

This is an overview of the SWAFCA still, a south view showing the arrangement of the condenser, distillation columns, cooking-fermentation tanks and steam furnace.

On top of each cooking-fermentation tank is a metal hatch. These hatches are shut during distillation and opened when cooking mash or cleaning the tanks.

sheet metal, angle iron and hinges. These hatches are shut during distillation, opened as necessary when cooking the mash or cleaning the tanks, and covered with burlap during fermentation. The burlap allows carbon dioxide, a by-product of alcohol (ethanol) production, to escape. At the same time, the burlap prevents the entrance of dirt, leaves or other foreign matter.

Once distillation has begun, alcohol vapors escape from the tanks through sheet metal ductwork, leading to the tall, central distillation column. These ducts are welded together. If only the central tank is used, the ducts from the two outer tanks may be blocked off by turning butterfly gate valves mounted within the ducts.

Inside the ducts mounted above the central tank, there's an automobile radiator with a copper line leading directly to the main, three-inch steam line. By turning a small valve, one can inject steam into the radiator to boost the temperature of the alcohol vapors just before they enter the distillation column. In fact, the extra heat may be needed to keep the alcohol vaporized: by this time, the vapors have traveled quite a distance from their primary source of heat, the steam furnace.

At the base of the distillation column on the furnace, and on each of the three tanks, there are combination temperature-pressure gauges. Pressure is not needed to process the corn mash through the cooking, fermentation or distillation stages. But, by providing both pressure and temperature in-

106

Mounted inside ductwork connecting the tanks to the distillation column is an automobile radiator. By opening small valve at left, still operators may inject steam into a radiator mounted inside ductwork. This boosts vapor temperatures.

dications, these gauges permit careful monitoring of the alcohol production processes, and the gauges warn of any possibly dangerous pressure build-ups.

Distillation Column. The 10-foot-high distillation column on the Selma plant is similar in purpose, but not in design, to most conventional, cylindrical distillation columns. At the Selma plant, ascending alcohol-water vapors pass through 40 horizontally-mounted, galvanized steel plates, each perforated with tiny, ¼-inch holes. As the vapors repeatedly come in contact with the 30-inch diameter plates, water remains on the plates or descends, while most of the alcohol remains vaporized at temperatures 173°F. or higher and continues upward to the condenser. As the alcohol progresses, it becomes increasingly enriched, sometimes reaching 185 proof at the top of the column. Column temperatures are carefully monitored with three thermometers, and, in the upper part of the column, temperatures are kept low enough so the water doesn't vaporize and continue to the condenser with the alcohol.

Conventional column distillation of alcohol is usually more complex. In the first place, most systems have either tanks or equipment* designed

---

*To separate solids from the liquid, one can simply let the contents settle for a day or two and draw the liquid off the top. But this takes time during which the mash may sour. Other methods are centrifuges, roller presses, and rotating screen drums.

to separate the mash liquid from solids. And this means that the liquid to be separated in the column enters the column as still beer. At Selma, the mash, including solids, is heated and the resulting vapors go directly into the column. More conventional distillation systems often have two columns rather than one. The second is sometimes referred to as a *rectifying column* or, more simply, as a *rectifier*. But its purpose is the same as the first — to distill or further separate alcohol from water. The use of two columns rather than one reduces the height of the system, making it easier to place in a barn or plant building.

There are other features that distinguish most conventional distillation columns from the one fabricated by the farmers' co-op in Selma. Most cylindrical columns are arranged so the still beer enters the upper third of the column and live steam, used as a heat source, enters the bottom. As the beer slowly descends through the perforated plates, the steam moves upward, stripping alcohol from the descending liquid. (*Stripping* means steam-heating the descending liquid enough so that alcohol vaporizes and moves on up the column.) The layers and layers of perforated plates permit multiple contacting of ascending steam and descending alcohol-rich liquid, causing this stripping process to be repeated many times. Steam may continue toward the top of the column with the alcohol vapors, but temperatures near the top are kept warm enough only for alcohol vapori-

The tall, white distillation column has horizontally mounted perforated plates inside that help separate alcohol vapors from ascending steam.

zation. Ideally, then, only alcohol-rich vapors continue to the condenser, while the steam reliquefies on the plates and then descends.

In most columns, water makes this descent through what are known as *downcomer pipes*. Typically, these pipes are about 1½-inches in diameter and about 6 or 7 inches long and are arranged vertically on alternating sides of the column. In a way, the pipes serve as a kind of small cesspool, draining off excess water that collects on one plate and passing it down to another. At the first, or upper, of these two plates, the pipe sticks up about ¾ inch above the plate surface. Once water on the plate reaches the ¾-inch level (as the result of excessive flow rates or plugging of the perforations), it begins to drain down the pipe to the plate below. The pipe descends to within ¾ inch of the next plate. A tiny, ¾-inch-high cup sits on the lower plate, just below the downcomer pipe. This cup forms a liquid "seal" on the bottom of the downcomer pipe. Without it, the heated vapor might travel through this passage as a shortcut of least resistance, rather than pass through the holes in the plates.

Condenser. After reaching nearly the top of the distillation column, the alcohol vapors travel horizontally out into the condenser. The basic parts of the condenser are: an eight-foot long, heavy metal tube-like shell, with a diameter of 10 inches; several feet of 1-inch-diameter copper tub-

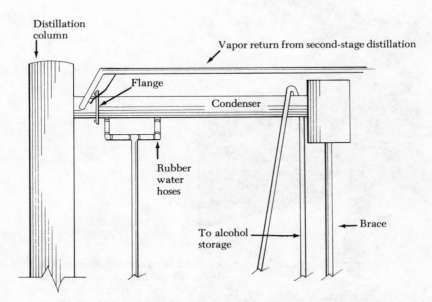

The condenser is mounted horizontally near the top of the distillation column.

ing; rubber hoses to bring cooling water into the condenser shell and to remove this water; and rubber hoses from the secondary distillation column and to the storage tanks. The vapors pass into the copper tubing, surrounded by water approximately 60°F. or colder. There, the vapors liquefy and return to the storage tanks as alcohol.

Although copper is used for the Selma condenser tubing and although most condenser tubing is made of copper alloys, aluminum, nickel alloys or chromium-nickel steels are also used. Usually, the tubing has an outside diameter of ½ inch to 1 inch. The tube sheets, into which the tubes are welded, expanded or packed, are generally of Muntz metal, naval brass, copper-silicon alloy or other corrosion-resistant alloys. However, according to some technical experts, the tendency of water to corrode metal surfaces is usually not a serious problem. Thin-walled heat-transfer surfaces or tubing of the less costly corrosion-resistant alloys are generally satisfactory.

There are a few other significant features to the condenser used at the Selma plant: Its shell is not connected directly to the distillation column; rather, it is attached with a normal pipe flange and rubber gasket. And, the flange connects the shell to another small, same-diameter tube. This

The condenser is secured to a flange that permits easy disassembly. Rubber hoses bring cooling water to the condenser.

smaller, six-inch-long piece is welded to the distillation column. The flange connection permits the detaching or disassembling of the shell for cleaning or repairs. Once returned to the desired condition, the shell is simply bolted back to the column at the flange. The gasket is an ordinary piece of rubber, cut to fit the tube circumference.

A two-inch, heavy rubber hose, attached at the base of the shell with a clamp-type fitting, supplies water to the condenser. A similar hose, with a shut-off valve, also pipes in alcohol vapors from the secondary distillation column that need recondensing. At end of the condenser shell farthest from the distillation column, there's a small end box, with another pipe at its base, used to carry condensate to the alcohol storage tanks. Both the condenser and the distillation columns are supported by metal braces, welded to and extending from either the combination tanks or the warehouse walls.

Secondary Distillation and Storage. Once alcohol vapors pass through the condenser and reliquefy, they then travel via a two-inch pipe toward a box-shaped storage tank. The 1,000-gallon tank is welded to stilts and elevated about eight feet above ground. A smaller line drops down from the two-inch pipe to a small sampling device. The device permits an operator to draw off a precise amount of alcohol to be tested for proof. With

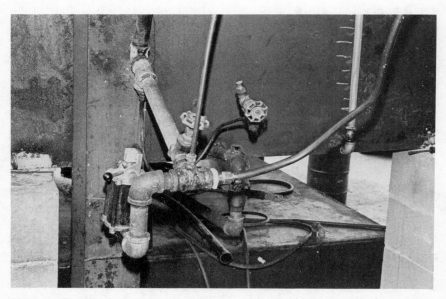

This sampling device allows still operators to test the alcohol proof. If necessary, the alcohol can be recycled to raise the proof to 160, the minimum fuel grade.

If the proof is inadequate, the alcohol is recycled through this 12-inch diameter, second-stage distillation column.

care and patience, Turner maintains, he can achieve up to 185 proof "on the first run." If the proof is adequate, the alcohol is diverted to a second, lower tank, where it is denatured with methanol and stored until used.

If the alcohol proof is too low, selected valves are used to divert the fuel to the *upper* storage tank. From this tank, the alcohol feeds by gravity to a second smaller, steam-heated, 12-inch-diameter distillation column. After redistillation, the alcohol vapors return back to the entrance end of the condenser. Here, the vapors recondense and then reenter the cycle once again; this time, the reliquefied alcohol should have a high-enough proof to be diverted directly to the lower tank. If large quantities of alcohol are produced, a third, box-shaped tank provides extra storage capacity. Each storage tank has a sight gauge for checking alcohol levels.

Cooking

The cooking, fermentation and distillation processes used by Turner and members of the South West Alabama Farmers' Cooperative Association (SWAFCA) do not typify most alcohol-producing operations. That is not to say they are unworkable for small-scale producers. There is a wide variety of cooking, fermentation and distillation methods, many of them workable, but most have distinguishing advantages or drawbacks. Once

This is a view looking down into a cooking-fermentation tank through an open hatch. Note the pipe grid and "legs" used to distribute steam.

again, the most noticeable feature of the SWAFCA plant is that the mash cooking, fermenting and heating-for-distillation steps all occur within the same three, 5,000-gallon tanks.

To begin, corn is machine-ground into a fine-to-grainy consistency; then it is fed into a pickup truck. In the meantime, the 5,000-gallon tanks are half filled with water. A fire is started beneath the steam furnace. Steam builds up, travels into the three tanks, sterilizes the tanks and agitates the water. After sterilization, the truckload of ground corn is brought to the edge of the tanks and 2,000 pounds of corn is slowly augured into each of the three tanks. Steam continues to increase the mixture temperature.

Once all 6,000 pounds of corn are augured into the three tanks, Turner adds 100 pounds of dolomite lime to each of the three tanks. This increases the alkalinity of the solution. For Turner's process, the optimum pH is 7. After he pours the lime into each tank, he tests the solution to ensure that it has a pH of 7. If necessary, he adds additional lime to increase alkalinity or sulfuric acid to reduce alkalinity, increase acidity and bring the pH down to 7. Turner adds pure, undiluted acid at a rate of one gallon at a time, then retests to ascertain whether additional acid is needed.

With the pH at 7, he adds one-half pint of the liquefying amylase enzyme *Tenase*, supplied by Miles Laboratories. The temperature of the solution is then increased to 200°F., at which point Turner adds a full pint

113

of *Tenase* and holds temperatures to 200° to 212°F. for 30 minutes. Then he reduces temperatures to 185°F., by adding cold water, and holds this temperature for 15 minutes.

Next, Turner tests the solution for sugar, using a hydrometer. If sugar is not indicated, he holds the tank mixture temperature at 185°F. for 20 minutes, then retests. If sugar is still not indicated, another one-half pint of *Tenase* is added and temperatures are held at 185°F. for another 15 minutes.

Ideally, the sugar level at this point would be about 15 percent, Turner said, adding that this level is high enough to achieve 10 percent alcohol when the solution is distilled. Once the sugar level is adequate, Turner cools the solution again, down to 140°F. Again, cold water is the cooling medium. Once 140°F. is reached, he adds two pints of a second enzyme, a sacchrifying enzyme known as *Diazyme*. The mixture then cools to 100°F., and finally a prepared yeast mixture (2 to 5 pounds of *saccharomyces cerevisiae* per ton of corn and several gallons of warm water) is added to each tank. Each tank hatch is left open, but covered with a gunny sack so dirt can't enter and heat can't easily escape, but carbon dioxide gases can.

Fermentation and Distillation

After the yeast is applied to each of the three tanks, fermentation proceeds. With warm temperatures, fermentation should be complete within 72 hours. In cold seasons, fermentation sometimes continues for more than a week. Ideal fermentation temperatures are about 85°F. Agitation is needed to expose the yeast to different parts of the mash throughout fermentation. Turner uses an air hose for agitation, similar to the air hoses at any automobile service station.

The discontinuation of gaseous bubbling on the mash surface indicates that fermentation is complete. At this point, the steam furnace is refired. Temperatures in the tanks are brought up to about 200°F., and the resulting alcohol-water vapors begin to proceed upward into the distillation column. As the vapors ascend, they begin to cool. The heavier water vapors remain on the column's perforated plates, while the alcohol vapors continue on to the condenser. Ideally, the temperatures at the top of the column are about 173°F., or the vaporization point for alcohol. Careful monitoring of temperatures, Turner says, is essential for effective, efficient distillation.

With distillation complete — a process that takes about four or five hours when all three tanks are used — the mash is allowed to cool. Once cooled, the mash is drained through 10-inch diameter holes on the south

Once the mash has cooled, it is drained through 10-inch pipes on the south sides of each of the three cooking-fermentation tanks.

end of each tank. Through the cooking processes, the used mash has become a high-protein feed that is fed to hogs and other livestock. During the cooking steps, the portholes are sealed with plates that screw into place. Rubber gaskets maintain a tight seal.

Turner and the members of SWAFCA rightfully call themselves pioneers in the field of small-scale, homemade alcohols. Whether SWAFCA gains its financial independence from the distillation plant remains to be seen. But the means for SWAFCA to lessen its dependence on traditional fuels is in place.

CHAPTER 7

Engine Modifications

To test the fuel produced from my 55-gallon still, I sought help from the president of a South Burlington car dealership known as The Automaster. And I got it, immediately and enthusiastically, from Jack Dubrul, who also happens to be a former automobile racer and engine mechanic. He was just the person for this project, although engine modifications for alcohol fuel are simple enough that a person with little mechanical skill can do them.

Within a few weeks, Dubrul repainted, tested and completed mileage checks for a 1974 Honda Civic. He also gave the four-cylinder car, with 57,840 miles on the odometer, a good tune-up. Then he invited me to come over to discuss the car modifications we'd undertake.

He outlined his plan while seated in his office, a square room with deep red, wall-to-wall carpeting, an intercom for quick calls to "service," and a seaman's porthole in one wall for receiving messages from secretaries and associates. Mementos of Dubrul's car racing and hot-air balloon-flying days are placed neatly about the room.

What he described sounded simple and inexpensive. To convert the Honda, or most any other car to operate on 160-proof alcohol (80 percent alcohol, 20 percent water), there is one mandatory modification: because the heat values for alcohol fuel are lower than for gasoline, the flow of fuel to the engine must be increased. The small, brass, carburetor fuel jets must be enlarged.

Before You Start

Legal Restrictions. Before disassembling your carburetor, check the clean-air laws of your state. There may be restrictions on the alteration of antipollution control devices or fuel systems. It is illegal for most automobile service stations to tamper with these devices and systems. Usually it is not illegal for an individual to do the same. But California, for example,

Jack Dubrul at the wheel of the 1974 Honda, powered by 160-proof alcohol fuel. Dubrul modified the car as an experiment for Garden Way Publishing.

is so strict in its enforcement of antipollution laws that waivers had to be obtained before the state's own experimental, alcohol-powered cars could be registered.

Check Warranty. Especially if you own a new car, check your manufacturer's warranty. General Motors, Ford and Chrysler say their warranties are not invalidated by the use of *gasohol*, a blend of 90 percent unleaded gasoline and only 10 percent 200-proof, anhydrous (water-free) alcohol. But a new car warranty may not remain in effect if you use, say, 160-proof alcohol in your engine.

Some other preliminaries: Obtain a drawing, or ask a mechanic for a manual, showing the parts of your carburetor. It might be a good idea to discuss the whole project with a mechanic, as I did. A carburetor rebuilding kit might also be helpful, but it is not essential.

Obtain a manufacturer's parts list if you need help understanding the carburetor assembly and operation. Also, rather than enlarging your jets with a drill, it may be easier simply to replace them with larger jets. Many jets are not straight; they have tapered or shaped nozzles that cannot be enlarged with a single drill. In any case, buy some extra fuel jets.

Tools. Most of the tools needed for carburetor conversion are simple to use, but you may not have them handy. Again, a good garage might be a source of help. A pair of needlenose pliers, a set of end wrenches, a screwdriver, and a small *hand* drill or brace for tiny bits are among the items needed.

The bits should obviously be the right size for the necessary enlargement. For the Honda, the secondary jet was enlarged with a 0.093 (No. 45) bit, and the primary jet was enlarged with a 0.052 (No. 55) bit. The size of your jet openings will most likely be different.

I would *not* recommend that you use a power-driven drill for enlarging the jets. The jets are so small and the material so soft that it can be done by hand.

Enlarging the Jets

Mechanics and fuel experts agree that the carburetor fuel jet or jets (if the carburetor is a multithroated model) must be enlarged about 40 percent to increase the flow of alcohol fuel to the engine. There are many different ways to proceed, and it's difficult to generalize for all the various makes of carburetors. But there are some general steps to follow that do apply. First, here's some more detailed information on how to select the right drill for your carburetor.

There are formulas for determining the bit size necessary for a 40 percent enlargement, but the simplest technique is to multiply the jet orifice by 1.4. For example, to go back to the Honda, the primary jet was enlarged with a 0.093 (No. 45) bit. To get this figure, the original diameter, 0.067,* was multiplied by 1.4 as follows:

$$0.067 \times 1.4 = 0.093$$

The same steps were followed for the secondary jet, and you could do the same with your own carburetor. Another more time-consuming but suitable procedure would be to simply increase the bit size by what you believe to be appropriate increments, and then retest the fuel each step of the way. If your carburetor is equipped with air jets, it may be necessary to enlarge these, to get the right mixture of fuel and air.

Carburetor Disassembly and Jet Drilling

Because there are so many different carburetors for domestic and foreign cars, I won't go into too much detail about how to disassemble one. As a first step, you might obtain or borrow a manual with blown-up illus-

---

*Remember, the numbers on jets are not always diameters. They may be part numbers. If you are in doubt, check the size of the orifice by noting the size of a bit that fits neatly into it.

trations of your carburetor. If you think it might be helpful, place all parts as you take the carburetor apart in small jars or envelopes that you can label.

Here's a suggested sequence for carburetor disassembly:

1. Remove the carburetor air filter housing, and all its hoses, and so on, from the engine.

2. Disconnect throttle linkages and, if appropriate, the manual choke.

3. With all hoses and attachments disconnected, remove the carburetor from the engine manifold.

4. Drain the carburetor and clean it.

5. Remove the carburetor air horn; if necessary, disconnect any linkages or rods.

6. Locate the jets; if necessary, remove any restrictions over them.

7. Remove the jets from either the main well or the main float bowl.

Dubrul prepares to remove the Honda air filter, one of the first obvious steps.

Once the air filter has been removed, the carburetor is exposed. This also must be removed.

The carburetor jets must be exposed and then removed for enlargement. One of the jets is near the tip of the screwdriver in this photo.

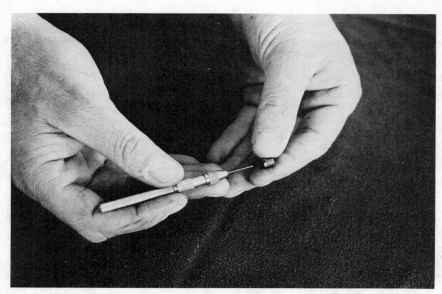

Use a hand drill to enlarge the small brass jets. Unconfirmed reports indicate that there's a danger the jets might explode if an electric drill is used.

Jets are round brass fittings with holes in their centers and slots in their tops. They are threaded, and they screw into place with a screwdriver placed in the slots.

8. Use the drill to enlarge the jets; keep track of primary and secondary jets to avoid replacing a jet in the wrong hole.

9. If necessary, enlarge the well diameters and passages to the jets, so all parts have the same diameter.

10. Replace the jets. Reassemble the carburetor.

After completing these steps for the Honda and pouring 160-proof alcohol into the fuel tank, we were reasonably pleased by the results. There was some hesitation. There was some difficulty starting the car on cold mornings. However, by opening up the manual choke when first operating the car, and by advancing the timing, we eliminated the hesitation.

Cold Weather Starting Device

With about $25 in parts, Dubrul fashioned a simple cold-starting device that enables easy starts from within the car. For those who don't

A starting device injects gas from a plastic reservoir into carburetor.

Close-up of starting device parts: reservoir (top left), fuel pump (top right), and neoprene gasoline line (lower left).

want to go to the trouble of building this device, an alcohol-fueled car may be started simply by pouring a cupful of gasoline directly into the carburetor. This is just enough for ignition. But it's a lot simpler, in the long run, to make yourself a starting device.

The device has these basic parts: a spring-loaded push botton, mounted on the dashboard and connected to a gasoline line; a windshield washer reservoir for the gas; neoprene gasoline lines; an electric fuel pump; and a nozzle. A push on the button squirts gasoline from the reservoir directly into the carburetor. It's that simple.

Air Preheating

A slightly more advanced modification, (and one that we have not found necessary to use) is an air preheating system. While gasoline will vaporize in the carburetor at freezing temperatures and remain in the

123

vapor state until it reaches the cylinder, alcohol becomes more difficult to vaporize as the thermometer goes below 50°F.

Even when vaporized in the carburetor, alcohol may condense on the cold walls of the intake manifold. Alcohol, compared with gasoline, produces larger droplets and rapid coagulation that makes proper distribution of the fuel-air mixture to each cylinder more difficult. This is the reason for the gas starter, and for the addition of air preheating. Air preheating is accomplished by passing the incoming air over part of the exhaust system. Light sheet metal scoops are the usual solution, while flexible metal ducting, of a diameter somewhat over that of the carburetor inlet, can also be run against a section of exhaust pipe.

The best place for an air intake heater is usually along the short but hot section of the exhaust manifold. Most air cleaner intakes can be rotated, and an inspection of your exhaust system will indicate that a three-sided

Dale Fillion, The Automaster service manager, pours 160-proof alcohol into the Honda fuel tank.

channel of sheet metal or a piece of flexible metallic duct, leading air *back*wards from the radiator, over the exhaust pipe, can be connected to the air cleaner horn (intake). These connections, and the fit of the scoop over the exhaust pipe up to the air cleaner, need not be perfect. The tuning of the carburetor, with and without the air cleaner, will be different.

Compression Ratio

Raising the compression ratio, while not essential, can improve performance and economy for cars operating on alcohol fuel. For this job, I'd recommend that you seek the advice of an experienced mechanic or machinist.

Driving With Alcohol

Although tests indicate that using alcohol does not cause damage to engine parts or components, some of the major automakers are advising caution. They suggest that it's too early to know precisely the long-term impact of using alcohol in a combustion engine. We have not noticed any significant damage to the 1974 Honda we are testing in Vermont. However, you can expect that alcohol will affect plastic parts.

This is the 1974 Honda, operating on 160-proof alcohol fuel.

All rubber or plastic components in the Honda's fuel system, such as the carburetor float, fuel filter, and fuel lines, were presoaked in alcohol. All survived except the fuel filter, which started to come apart. The filter was replaced with a metallic type that survived the soaking. *Viton*, a synthetic material often used for the fuel inlet needles, will swell when exposed to alcohol. Again, steel replacements may be necessary.

Alcohol also acts as a kind of cleaning solvent for some coatings and gasket cements. Therefore, all filters should be inspected frequently, at least daily during the first few tankfuls of alcohol-powered driving. Older fuel tanks build up a lacquer-type residue that the alcohol initially breaks loose, producing a gunk that clogs the filters. Just change the filter after the first or second tankful of alcohol fuel. There's no need to worry about permanent damage to the fuel tank.

Fuel tanks should be vented to avoid excessive pressure. If excessive pressure builds up, there may be a tendency of the alcohol to erupt or spurt out of the fuel tank when the cap is removed. This tendency is more pronounced during warm or hot days, or with fuel tanks located near warm engines. The cap should fit properly and the vent should be free and open. Watch out for nonstandard caps with bent or clogged vents. Alcohol should also be stored in containers with free and open vents.

CHAPTER 8

Federal Rules and Regulations

The federal Bureau of Alcohol, Tobacco and Firearms (BATF) administers Internal Revenue laws related to spirits or alcohol. These laws and many regulations have been modified so the BATF can administer the making of alcohol for fuel purposes. It is relatively simple for the small-scale experimenter to obtain a federal permit authorizing the making of alcohol. According to the Internal Revenue Code, everyone must apply for and receive a permit "before commencing the business of a distiller."

Two Types of Permits

The law provides for two types of distilled spirits plant (DSP) permits: First, the *operating* permit authorizes the production of alcohol for commercial sale or nonexperimental purposes. Applying for this permit is a complex, time-consuming process. The second, an *experimental* distilled spirits plant permit, authorizes experimentation and development of distillation and industrial uses of alcohol.

It is this second type of permit that small producers should apply for. The experimental permit does not authorize the sale of alcohol; it does provide for the use of alcohol by more than one person, provided that the alcohol is used on the premises described in the permit application. The alcohol may be used off premises if it is denatured. To apply for an experimental distilled spirits permit, simply send a letter (two copies) to the nearest regional BATF administrator, describing your still. (A list of BATF offices follows.) Here are the basic points of information, as stated by the bureau, that must be included in your letter to the BATF:

NATURE AND PURPOSE. Give a general statement describing what you intend to do. Example: Applicant wants to experiment with solar energy to produce ethyl alcohol which will be used as fuel for farm engines and heaters; including, but not limited to tractors, combines, swathers, cars, trucks, irrigation motors, houses, grain dryers, etc.; or, applicant intends to experiment with waste products (corn cobs, stalks, spoiled grain) to produce ethyl alcohol to determine if it can be refined and used as a fuel to run farm implements; or, simply, applicant intends to build a still which can efficiently refine ethyl alcohol produced from waste products for ultimate use as a fuel.

DESCRIPTION OF PLANT PREMISES. Describe the location of the premises where you intend to establish the experimental plant. If you are a farmer, this should include your entire farm to enable you to use the alcohol you produce without removing it from your "plant premises." Include in your description the number of acres involved. Also, describe the buildings used in the production and storage of alcohol (if applicable) and their relative location on your farm.

DESCRIPTION OF PRODUCTION PROCESS AND EQUIPMENT.   Describe the production process you intend to use. Example: "Mash to be fermented will consist of spoiled grain, vegetables, and kitchen garbage. Initial distillation of ethyl alcohol will be accomplished by solar energy. Ethyl alcohol recovered from the

	1. PERMIT NUMBER X-DSP-VT-6
DEPARTMENT OF THE TREASURY — BUREAU OF ALCOHOL, TOBACCO AND FIREARMS	
	2. EFFECTIVE DATE JAN 14 1980
EXPERIMENTAL DISTILLED SPIRITS PLANT PERMIT UNDER 26 U.S.C. 5312(b)	3. DATE OF APPLICATION October 24, 1979
	4. EXPIRATION DATE January 31, 1982

5. NAME OF PERMITTEE AND ADDRESS OF PERMIT PREMISES

Frederick W. Stetson
Colchester Pond Road
Colchester, Vermont 05446

"This permit is conditioned on compliance by you with the purposes of the Federal Water Pollution Control Act (33 U.S.C. 1341(a))."

THIS PERMIT IS GRANTED FOR EXPERIMENTATION OR DEVELOPMENT OF SOURCES OF MATERIALS FROM WHICH SPIRITS MAY BE PRODUCED; PROCESSES BY WHICH SPIRITS MAY BE PRODUCED OR REFINED; OR INDUSTRIAL USES OF SPIRITS.

6. LIMITATION

Pursuant to application and subject to applicable law and regulations and to the conditions set forth on the reverse side of this permit, you are hereby authorized and permitted to engage, at the above address, in the operations specified herein.

This permit will remain in force until suspended, revoked, voluntarily surrendered, or automatically terminated on the expiration date noted above.

This permit is not transferable. In the event of any lease, sale, or other transfer of the operations authorized, or of any other change in the proprietor of such operations, this permit shall automatically terminate. (If permittee is a corporation or partnership, see reverse side.)

7. SIGNATURE OF REGIONAL REGULATORY ADMINISTRATOR, BUREAU OF ALCOHOL, TOBACCO AND FIREARMS

ATF F 5110.65 (9-79)

This is the experimental distilled spirits permit issued for the author's stills.

mash via solar still will be refined in a kettle still of my own manufacture."
Then give a descriptive list of the equipment used in the process. Include all
equipment from the mash tank to the finished alcohol storage tank. In addition, describe when and how you intend to denature the alcohol produced,
i.e. alcohol will be mixed with gasoline in a 10:1 ratio immediately after production. This will be done in the fuel storage tank.

SECURITY. Tell what security measures will be provided for the alcohol produced. Example: Alcohol will be stored in a 1,500-gallon fuel tank which is
equipped for locking with a padlock; or, alcohol will be immediately denatured, drummed off, then stored in a locked shed. All windows in the shed
are equipped with security screens and two watchdogs are on the premises.

RATE OF PRODUCTION. State in gallons, the amount of alcohol you expect to
produce in an average 15-day period. You may estimate if you are unsure of
the amount at this time. Give the average proof of the finished alcohol produced. Example: Approximately 400 gallons of alcohol, averaging between
160-190 proof will be produced in a 15-day period.

DISPOSAL. Describe how you will dispose of any waste water or by-products
from your still. Wastes should be disposed in a way that does not pose any
health hazards and there should be no chance of any harm to navigable
waters.

The "Revenuer's" Visit

After you send your letter to the BATF, expect a member of the
bureau's staff to contact you and ask to inspect your still, even if it is a
small one. I live in Vermont and I received a visit from a BATF inspector
from Boston. The nearest BATF *regional office*, however, is in New York.
The Boston inspector examined my still, asked a few simple questions,
took pictures and then told me, on the way out the door, that he would
recommend to the New York office that my application be approved.

He also described the BATF's regulation, requiring that I file a
distiller's bond with the BAFT regional office to cover the liability for
federal tax on the alcohol I might produce. I discovered later that I could
file a $100 *cash* or surety bond if I intended to produce not more than
2500 gallons of alcohol a year. The acceptance of cash bonds, in the form
of a certified check or postal money order, was another change in the
BATF rules to help the small distiller. The necessary forms for the bond
may be obtained from the nearest BATF regional office.

In early 1980, the BATF waived the $100 bond requirement and said
that regulations for *experimental* distillers apply to those producing up to
5000 gallons a year. Most experimental distillers who have submitted
bonds will have them returned, according to a BATF spokesman.

A permit or authorization issued by the BATF *does not relieve you of the responsibility for complying with any state and local requirements* concerning the production and use of ethyl alcohol. Check your local health, zoning or planning offices for information or assistance. Or, you could contact your regional BATF office if further assistance is needed.

The BATF Responds

Here are some commonly asked questions about experimental stills or distillation plants. Answers have been provided by BATF offices in Washington, D.C. and Chicago, Illinois:

QUESTION: Can I sell or loan any excess alcohol produced to another person for fuel use?

ANSWER: An experimental distiller may not produce alcohol principally for sale. Undenatured alcohol may be used as fuel only at the plant premises described in your bond and letter application. Only a plant qualified as a commercial DSP can sell, loan or give alcohol to another party.

QUESTION: Can I remove some of the alcohol from the plant premises for my own use? (*Example*: as fuel for my personal car).

ANSWER: You may remove alcohol from your plant premises for your own use as a fuel; however, the alcohol must be *completely* denatured according to one of the two formulas listed below before removal:

Formula No. 18. To every 100 gallons of ethyl alcohol add:

   2.5 gallons of methyl isobutyl keytone:

   0.125 gallon of pyronate or a compound similar thereto;

   0.50 gallon of acetaldol; and

   1 gallon of either kerosene or gasoline.

OR

Formula No. 19. To every 100 gallons of ethyl alcohol add:

   4.0 gallons of methyl isobutyl keytone;

   and

   1.0 gallon of either kerosene or gasoline.

QUESTION: Must I denature the alcohol before using it on my farm?

ANSWER: No. You may use undenatured alcohol anywhere on your plant premises. However, if you denature your alcohol with gasoline, diesel fuel or heating fuel immediately after production, we may approve less stringent security systems and recordkeeping requirements than are imposed on applicants who do not denature their alcohol.

QUESTION: Can any of the alcohol produced be used for beverage purposes?

ANSWER: Absolutely not. Besides the IRS Excise Tax of $10.50 per proof gallon, which you would become liable for, you could also incur severe criminal penalties.

QUESTION: Can I build my distillery system prior to receiving any authorization from BAFT to operate?

ANSWER: Yes. However, you must file an application to establish an experimental distilled spirits plant with BATF immediately after its completion. You may, however, file sooner if you feel you will have the equipment set up prior to the qualification visit by the BATF inspector. However, under no circumstances may you start fermenting mash or producing spirits prior to receipt of a formal authorization by the Bureau.

QUESTION: Can I qualify two or more farms for my plant premises?

ANSWER: Yes, if they are in close proximity to each other to allow a BATF inspector to inspect all premises without causing undue travel and administrative difficulties.

QUESTION: Can a group of five or 10 individuals cooperatively construct a plant and distribute the ethyl alcohol produced at the plant among themselves?

ANSWER: A group may file application for qualification of an experimental distilled spirits plant with a single production facility, and describe in the application the premises of each individual where the alcohol will be used. Alcohol produced at the plant may be used tax-free only on the premises described in the application. The distiller's bond must be conditioned to provide coverage for the alcohol while in transit between the production location and the various premises where the alcohol will be used.

If the alcohol will be completely denatured in accordance with a prescribed formula, the alcohol can be removed from the plant for use by cooperative members without additional BATF restrictions.

QUESTION: How long a period is the authorization effective?

ANSWER: We are currently approving operations for a two-year period unless you include in your application some justification for a longer period of time.

QUESTION: Can I renew my experimental plant authorization after expiration?

ANSWER: Yes. When the authorization expires, you may file a new application listing the current information on all subjects originally described (security, rate of production, equipment, etc.).

QUESTION: Can partnerships and corporations make application as well as individuals?

ANSWER: Yes, however, if the application is filed by a partnership, all partners must sign it. If it's filed by a corporation, a person authorized by the corporation must sign and proof of such authorization must accompany the application (e.g. certified copy of a corporate resolution or abstract of bylaws giving such authority).

QUESTION: Can I convert to a commercial operation?

ANSWER: There is no simple means of converting an experimental operation into a commercial operation. Normally, all provisions of Title 26, U.S.C. Chapter 51 and Title 27, CFR Part 201 will be waived for an experimental alcohol fuel-related DSP except for those relating to:

(1) Filing application for, and receiving approval to operate an experimental DSP for a limited, specified period of time;

(2) Filing of a surety bond to cover the tax on the alcohol produced;

(3) Attachment, assessment and collection of tax;

(4) Authorities of BATF Officers; and

(5) Maintenance of records.

No such blanket waiver will be given for commercial operations. You will have to follow the qualification procedure outlined in BATF P 5000.1. We will, however, give favorable consideration to alternate procedures from regulations which do not present a definite jeopardy to the revenue; however, each such variation will be viewed and ruled upon an individual basis.

Remember to check your nearest BATF office for any last-minute changes to regulations or distilled spirits permit requirements.

ADDRESSES OF THE REGIONAL OFFICES OF THE BUREAU OF ALCOHOL, TOBACCO, AND FIREARMS

Central Region

Indiana, Kentucky, Michigan, Ohio, West Virginia

Regional Regulatory Administrator
Bureau of Alcohol, Tobacco and Firearms
550 Main Street
Cincinnati, Ohio 45202
800-582-1880, Ohio; 800-543-1932, all other

Mid-Atlantic Region

Delaware, District of Columbia, Maryland, New Jersey, Pennsylvania, Virginia

Regional Regulatory Administrator
Bureau of Alcohol, Tobacco and Firearms
2 Penn Center Plaza, Room 360
Philadelphia, Pennsylvania 19102
800-462-0434, Pennsylvania; 800-523-0677, all other

Midwest Region

Illinois, Iowa, Kansas, Minnesota, Missouri, Nebraska, North Dakota, South Dakota, Wisconsin

Regional Regulatory Administrator
Bureau of Alcohol, Tobacco and Firearms
230 S. Dearborn Street
15th Floor
Chicago, Illinois 60604
800-572-3178, Illinois; 800-621-3211, all other

North-Atlantic Region

Connecticut, Maine Massachusetts, New Hampshire, New York, Rhode Island, Vermont, Puerto Rico, Virgin Islands

Regional Regulatory Administrator
Bureau of Alcohol, Tobacco and Firearms
6 World Trade Center, 6th Floor
(Mail: P.O. Box 15,
Church Street Station)
New York, New York 10008
800-442-8275, New York; 800-223-2162, all other

Southeast Region

Alabama, Florida, Georgia, Mississippi, North Carolina, South Carolina, Tennessee

Regional Regulatory Administrator
Bureau of Alcohol, Tobacco and Firearms
3835 Northeast Expressway
(Mail: P.O. Box 2994)
Atlanta, Georgia 30301
800-282-8878, Georgia; 800-241-3701, all other

Southwest Region

Arkansas, Colorado, Louisiana, New Mexico, Oklahoma, Texas, Wyoming

Regional Regulatory Administrator
Bureau of Alcohol, Tobacco and Firearms
Main Tower, Room 345
1200 Main Street
Dallas, Texas 75202
800-442-7251, Texas; 800-527-9380, all other

Western Region

Alaska, Arizona, California, Hawaii, Idaho, Montana, Nevada, Oregon, Utah, Washington

Regional Regulatory Administrator
Bureau of Alcohol, Tobacco and Firearms
525 Market Street
34th Floor
San Francisco, California 94105
800-792-9811, California; 800-227-3072, all other

Appendices

Suppliers for Small Still Users

Ametek U.S. Gauge Division 900 Clymer Avenue Sellersville, PA 18960 (215) 257-6531	*gauges, thermometers*
Arthur H. Thomas Co. P.O. Box 779 Philadelphia, PA 19105	*proof hydrometers*
Bacchanalia P. O. Box 1 Dartmouth, MA 02714 (617) 636-5154	*fermentation locks, pH test strips, plastic fermentation vessels, triple- scale hydrometers, proof hydrometers*
BIOCON (U.S.) INC. 261 Midland Avenue Lexington, KY 40507 (606) 254-0517 (606) 254-0518	*enzymes, yeasts*
Garden Way Catalog Charlotte, VT 05445 (802) 425-2121	*plastic buckets (6 gallon),* All- American *pressure canners, corn shellers, yeasts, floating thermometers, fermentation locks and corks, triple- scale hydrometers and test jars.*
Glen-Bel Enterprises Route 5 Crossville, TN 38555	*corn shellers*
H-B Instrument Co. American and Bristol Streets Philadelphia, PA 19140 (215) 329-9125	*proof hydrometers*

Industrial Innovators, Inc.
P. O. Box 387
Ashford, AL 36312
(205) 899-3314
(205) 899-3381

steel and copper farm stills

International Fuel Systems
1820 West 91st Place, Suite 100
Kansas City, MO 64114
(816) 361-9900

*100- to 1,000-gallon stills; produce
two to ten gallons per hour*

Micro-Tec Laboratories, Inc.
Route 2, Box 19
Logan, IA
(712) 644-2193

*enzymes, pH test papers, triple-scale
hydrometers, proof hydrometer,
books, plans*

Mother's Plans
P. O. Box A
Flat Rock, NC 28726
(800) 438-0238

*hydrometers, fermentation locks,
enzymes, still plans, books*

The Revenoor, Inc.
P. O. Box 185
La Center, WA 98629
(206) 263-2200

*stills capable of producing from less
than 1 to more than 300 gallons per
day*

Rutan Publishing
P. O. Box 3583
Minneapolis, MN 55403
(612) 874-7812

*hydrometers, proof hydrometers, pH
test papers, books, enzymes*

Still Company
115 E. 9th Street
P. O. Box 9570
Panama City Beach, FL 32407

stills with 12-quart boilers

Stills
Box 1493
Pine Bluff, AR 71613

small alcohol fuel kits

Victory Stills
3317 Tait Terrace
Norfolk, VA 23513
(804) 855-7110

*"experimental" 5-gallon stills;
developing 55-gallon stills*

Wisconsin Aluminum Foundry
PO Box 246
Manitiwoc, WI 54220

pressure canners

Corn Mash Recipes Using Novo and Miles Enzymes

MATERIALS FOR MASHING

1 triple scale wine hydrometer; measures sugar content, alcohol potential
and specific gravity of mash
1 proof hydrometer (0–200 proof)
1 long-handled wooden spoon
1 floating thermometer (0°–212° F.)
1 cheesecloth or burlap bag
½ cup of flour for sealing paste
1 measuring cup and measuring spoons
1 bottle iodine tincture
pH testing paper with a range of 3 to 9 or 4 to 8
6 mason jars or similar containers
1 vise-grip wrench or pliers
1 pair of rubber gloves

INGREDIENTS FOR MASHING*
WITH NOVO AND MILES ENZYMES

7 pounds of ground corn meal
2½ gallons of water
3-5 teaspoons of citric acid†
2 teaspoons of amylase enzyme (Thermamyl 60L) (Taka-Therm)
2 teaspoons of gluco-amylase enzyme (AMG 200L) (Diazyme)
1 package of distiller's yeast (*sacharomyces cerevisiae*); bakers yeast
may be substituted

---

* For larger batches of mash, these ingredients may be increased approximately propor-
tionately. For example, if you're mashing with a bushel of corn (56 pounds) in a 55-gallon
drum, you should use about 20 gallons of water, 10-12 teaspoons of amylase enzyme, 10-12
teaspoons of glucoamylase and 10-12 teaspoons of yeast.

† Hydrochloric or sulphuric acid may also be used diluted, 1 part acid to 9 parts water. If
used, exercise extreme caution; wear rubber gloves and protective glasses.

Ingredients for mashing with Novo products include, from left, corn, yeasts, citric acid (a substitute for sulphuric acid), and lime; in rear are Novo enzymes.

Mashing with Novo Enzymes

As before, the first step is to ensure that all containers, materials and implements are clean. Place your pressure cooker or cooking container on a stove or heat source and fill it with 2½ gallons of water, hot from the faucet. Slowly increase the temperature of the water until it reaches about 150° F. Then add the 7 pounds of corn and stir the mixture well. Continue to increase the temperature of the mash.

Adjusting pH

Dip your measuring cup into the pressure cooker and take a small sample of the mash (less than ¼ cup is sufficient) and place a small piece of pH testing paper in the sample. (If you place the paper directly into the cooker, you'll probably lose it.) The pieces of pH testing paper needn't be any longer than a new eraser at the end of a pencil. The pH is a measure of acidity or alkalinity of the solution; sometimes these are also called the solution's acidness or "sweetness." A pH count of 7 is neutral; below 7 is acid; above 7 is alkaline.

The pH of your mash will be indicated by the color of the testing paper. At this stage, a pH count of 6.5 is optimum for the enzyme, *Thermamyl 60 L*. Frankly, I think it's difficult at first to establish the exact color of the paper, but after a few tries it becomes easier, the colors are more ap-

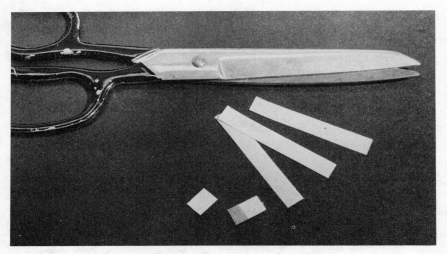

Cut pH test strips into small pieces for testing mash acidity.

parent, and matching the color of the paper with numbered color samples to determine pH is not impossible.

If necessary, adjust the pH of the mash until it has a count as close to 6.5 as possible. To raise the pH count, add agricultural lime, one tablespoonful at a time; to reduce, add citric acid, also in small, gradual amounts. Then retest with pH paper as often as is necessary. I've found that one or two adjustments usually bring the pH to a count close to the figure I'm looking for. (If you use sulfuric or battery acid to adjust the pH,

To establish pH, place a small sample of mash in a cup. Then drop test paper into cup.

dilute it first by adding one part acid to nine parts water. And use care. This strong acid *burns*. Wear rubber gloves. Wash with cold water immediately if you spill this acid on yourself. Be especially careful to avoid contact with your eyes and face. Use small amounts of diluted acid to make adjustments.)

Adding Enzymes

Once the solution is adjusted and thoroughly mixed, add two teaspoonfuls of the first enzyme, *Thermamyl 60 L*. This is a liquefying enzyme supplied by Novo Laboratories of Wilton, Connecticut.

After adding the enzyme, continue to stir the solution and gradually increase the mash temperature. Use your floating dairy thermometer to check temperatures, and wear rubber gloves. Plucking a thermometer from a near boiling mash with bare hands is no fun. Novo recommends that temperatures be increased to 221° F. (105° C.), to achieve the best enzyme activity. It may take about an hour to bring your mash to that temperature. And, if your stove works the way mine does, you may not be able to achieve mash temperatures much above boiling (212° F.). In any case, get the temperatures as high as possible without exceeding 221° F. Above 221° F., the enzyme activity and starch breakdown declines.

Hold the mash as close to 221° F. as possible for 60 to 90 minutes, while

A surprisingly small amount of enzymes is necessary to help convert starch to sugar.

continuing to stir. Then reduce or turn off your heat, and allow the mixture to cool to 140° F. (60° C.). You can cool the mixture by adding about a gallon of ice water, one quart at a time. Or, if additional cooling is needed, fill a plastic bag with ice cubes and place this in the solution. This method cools the mash without excessive dilution. Once again, it may be difficult to maintain the exact temperatures called for. I wouldn't worry about this; just try to adjust as necessary. For example, if you cool the mash and temperatures fall below 140° F., increase your holding time.

Once the mash temperature reaches about 140° F., it's time to again readjust the pH to prepare the mash for a dose of a second enzyme, *AMG 150 L.* This is a "saccharifying enzyme," also known as a gluco-amylase, that breaks down the starches further and completes their conversion to fermentable sugars. Adjust the mash by adding acid as necessary to achieve a pH count of about 4.5. Once achieved, add two teaspoonfuls of AMG 150 L. Hold the mash at 140° F. for one hour. Continue to stir.

To achieve optimum conversion during mashing, try to ensure that the mash pH count does not fall much below 4.5 or increase above 5. To do this, recheck the mash pH after 30 minutes of holding at 140° F. Unwanted bacterial contaminants grow rapidly when the pH count is above 5; but, if the count falls below 4, the solution becomes so acidic that other desirable chemical processes are inhibited.

Cooling the Mash

Turn off all heat beneath your cooker and allow the mash to cool below 90° F., but, if possible, no lower than 70° F. Fermentation will be inhibited below 70° F. and stopped altogether at temperatures above 90° F.

Testing for Sugar

After cooling the mash, a second test should be made to determine its sugar content. For optimum fermentation, sugar should be at an optimum level, about 10 to 15 percent. To determine the sugar content of your mash, use a triple-scale wine hydrometer with a Brix or balling scale. This scale indicates sugar content by weight.

It is also helpful, but not essential, to use your hydrometer to take a reading of *potential alcohol*. By doing this, you'll know the approximate potential of your mash to produce the desired end product: alcohol fuel.

Fermentation

After carefully converting the corn starch to sugar, and cooling your mash to fermentation temperatures (85°-90° F. is optimum; 70°-90° F. is acceptable; 60°-90° F. will work), your next step is to convert the sugar to alcohol. A simple way to ferment your mash is to pour it into a large plastic container with a tightly fitting lid. Then puncture the lid and place a fermentation lock into the hole.

To get the fermentation going, dump the contents of one packet* of yeast into the mash, then mix the solution slowly and thoroughly. Mix again, about once every two or three hours.

The final step prior to distillation is to separate the solid and liquid parts of the mash.

Distillation

Distill liquid "beer," as explained in Chapter 2, pages 27–34.

BASIC STEPS IN ALCOHOL PRODUCTION,
USING MILES LABORATORIES ENZYMES
IN A 5-GALLON STILL

1.	Preparation	Ensure that all containers are clean. Collect all materials, including corn meal or grains ground to a fine meal (12-16 mesh). Wheat, rye, sorghum and barley are some of the other suitable grains. The following recipe is for corn meal.
2.	Slurrying	Add 2½ gallons of water to 7 pounds of corn and increase heat. Adjust mash to a pH of 6.0 to 6.5. Mix well.
3.	Cooking and Liquefaction	Add 2 teaspoons of *Taka-Therm*, the liquefying enzyme. Continue to increase heat to 212° F. Mix constantly. Increase heat slowly.
4.	Conversion (Saccharification)	Cool the mash to 135° to 140° F. Adjust pH to 4.0 to 4.5. Add 2 teaspoons of *Diazyme*, the gluco-amylase enzyme. Cool mash as quickly as possible to fermentation temperature (85° to 90° F.).
5.	Fermentation	Add 2 teaspoons of distiller's yeast (one packet). Let stand for 48 to 120 hours.
6.	Distillation	Separate mash solids and liquids. Distill liquid "beer."

---

*One packet of dry brewer's yeast weighs 7 grams.

Consultants and Suppliers for Farm and Large Still Users

Consultants for Farm and Large Still Users

The Alcohol Technology Corp.
3191 "M" St.
Box 2365
Merced, CA 95340
(209) 383-3147

Harland Anderson
Five Woodcrest Drive
Burnsville, MN 55337
(612) 825-9451

Daniel Archer
1351 Waconia Avenue, SW
Box 1445
Cedar Rapids, IA 52406
(319) 398-0644

William P. Bailey
2508 Northwest 87th
Seattle, WA 98117

Ernie Barcell
Seven Energy Corp.
3760 Vance
Wheat Ridge, CO 80030
(303) 425-4239

Bartlesville Energy
  Technology Center
Bartlesville, OK 74003
Contact: Jerry Allsup
(918) 336-4268

Battelle Columbia Laboratories
505 King Avenue
Columbus, OH 43201
Contact: Bill Allen
(614) 424-6424

Ulrich Bonne
Honeywell Inc.
Corp. Tech. Center
10701 Lyndale Avenue South
Bloomington, MN 55420
(612) 887-4477

Glen Brandt
Brandt Chemical Co.
P. O. Box 277
Pleasant Plains, IL 62677
(217) 626-1123

Center for Biology of Natural
  Systems
Washington University
St. Louis, MO
Contact: David Freedman
(314) 889-5317

Robert Chambers
808 South Lincoln #14
Urbana, IL 61801
(217) 384-8003

Miles Connors
One Penn Plaza
New York, NY 10001
(212) 736-1500

Development Planning and
  Research Associates
2000 Research Drive
P. O. Box 727
Manhattan, KS 66502
Contacts: Milton David,
  Robert J. Buzenberg
(913) 539-3565

Dale Devermon
3550 Great Northern Avenue
Route 4
Springfield, IL 62707
(217) 787-9870

EG&G Idaho, Inc.
P. O. Box 1625
Idaho Falls, ID 83415
Contact: Don LaRue
(208) 526-0509

Edward Falck & Co.
6125 Eye St., N.W.
Washington, D.C. 20006
(202) 331-1989

Enerco, Inc.
139A Old Oxford Valley Road
Langhorne, PA 19047
Contact: Miles J. Thomson
(215) 493-6565

Energy Inc.
P. O. Box 736
Idaho Falls, ID 83401
Contact: Steve Winston
(208) 524-1000

Environmental Group
RD #3
Quakertown, PA 18951
Contacts: Jack Hershey,
  Bob Meskunas
(215) 536-8243

Galusha, Higgins and Galusha
P. O. Box 751
Glascow, MT 59230
Contact: Jim Smreka
(406) 228-9391

Gasohol Production, Ltd.
174 New Mark Esplanade
Rockville, MD 20850
(301) 251-9279

Joe Glasset
2191 N. 1400 E.
Provo, UT 84601
(801) 378-6242

Joseph L. Gordon
Box 7808
Boise, ID 83729
(208) 386-5670

William S. Hedrick
844 Clarkson
Denver, CO 80218
(303) 832-1407

Pincas Jawetz
Independent Consultant on
  Energy Policy
425 East 72nd Street
New York, NY 10021
(212) 535-2734

Dick Johnson
Felox Corporation
7703 Normandale Road
Minneapolis, MN 55435
(612) 835-1103

William J. Jones
1818 Market Street
Philadelphia, PA 19103
(215) 299-8193

E. Kirchner
10 South Riverside Plaza
Chicago, IL 60606
(312) 454-3685

Jim Kumana
1050 Delta Avenue
Cincinnati, OH 45208
(513) 871-7500

Midwest Solvents Co., Inc.
1300 Main Street
Atchison, KS 66002
(913) 367-1480

James Miles
Box 83, Route 1
New Albin, IA 52160
(507) 724-2387

Donald Miller
2900 Vernon Place
Cincinnati, OH 45219
(513) 281-2800

Dr. Ing. Hans Mueller
Chemapec, Inc.
230 Crosways Park Drive
Woodbury, NY 11797
(516) 364-2100

Marvin Oerke
RFD 3, Box 194
Butler, MO 64730
(816) 669-5159

Jim Pufahl
Box 99, Route 2
Milbank, SD 57252
(605) 432-4169

Lloyd Reeser
Route 1
Weldon, IL 61882
(217) 736-2539

LeRoy Schartz
RFD 3
Great Bend, KS 67530
(316) 793-7144

Dr. L. Eugene Schroder
North Route
Compo, CO 81029
(303) 523-6787

Eldon L. Shelter
S&S Galvanizing Co.
P. O. Box 37
Clay Center, NE 68933

Don Smith
Rural Route 1, 650 Pine
Colby, KS 67701
(913) 462-7531

Biomass Suchem
Clewiston, FL 33440
Contact: Dr. Ron DeSpephano
(813) 983-8121

Roger Sweet
Agrifuels, Inc.
Crookston, MN 56716

TRW Energy Systems Group
8301 Greensboro Drive
McLean, VA 22102
Contacts: V. Daniel Hunt
          Mani Balasubramaniam
          Harlan L. Watson
          Warren Standley
(703) 734-6554

Albert Turner
Southwest Alabama Farmers
   Co-op
Selma, AL 36701
(205) 683-8800

Jackson Yu
50 Beale Street
San Francisco, CA 94119
(415) 768-2971

Suppliers for Farm and Large Still Users

ACR Process Corp.
602 East Green St.
Champaign, IL 68120
(217) 351-7510

technology, design consulting and training; plant size: 40,000 to 1.5 million gallons per year, and larger

Abbeon Industrial Lab and Plant
123-52 A Gray Avenue
Santa Barbara, CA 93101

components

Agri-Fuel Corp.
200 Market Building
Suite 961
Portland, OR 97201
(503) 223-6660

specializes in continuous-flow fermentation processes and equipment, with a minimum annual capacity of 250,000 gallons

Agri-Hol International
2000 W. 98th Street
Bloomington, MN
(612) 888-0018

sells a unit built by Winnebago with an annual capacity of 500,000 gallons

Agri Stills of America
3550 Great Northern Ave. RR#4
Springfield, IL 62707
(217) 787-4233

manufacturer and distributor of alcohol plants and equipment (plants have 250 bushel capacity)

Agri-Systems International
2301 Independence Blvd.
Kansas City, MO
(816) 231-6990

integrates alcohol fuel systems for farms

Agrodyne
P. O. Box 934
Idaho Falls, ID 83401
(208) 524-1000

development, promotion, and implementation of ethanol production equipment; supplies; complete services; plants with 200,000 to 300,000 gallons per year capacity

ALCOGAS
220 Equitable Building
730 17th St.
Denver, CO 80203
(303) 572-8300

design, consulting, technology, engineering; plant sizes: 55,000 to 2.5 million gallons per year

Alcohol Fuels, Inc.
Franklin, NE 68939
Contact: Brian Hayes
(303) 425-3101

farm plants (125 gallons per hour), enzymes

Alcohol Technology Corp.
3191 "M" Street
Box 2365
Merced, CA 95340
(209) 383-3147

technical and consulting services; blue prints for fermentation and distillation equipment.

Arlon Industries, Inc.
P. O. Box 347
Sheldon, IA 51201
(712) 324-3305

distillation column and equipment manufacturer

Bechtel Corp.
(C & I Girdler Co.)
1721 South 7th Street
P. O. Box 32940
Louisville, KY 40232
(502) 637-8701

commercial still engineering

Brannon, Camp and Wimbish
6755 Peachtree Industrial Blvd.
Suite 108
Atlanta, GA 30360
(404) 449-3333
(404) 449-8282

design and process engineering for alcohol fuel plants; still columns

Brown-McKee, Inc.
P. O. Box 2878
906 Slaton Rd.
Lubbock, TX 79408
(806) 745-4511

plant contractors and engineers

Bush-Bohler Brothers of America, Inc.
1623 West Belt North
Houston, TX 77043
(713) 465-8376

technology, design, engineering, consulting, construction; plant size: 500,000 to 100 million gallons per year

CCI Industries
27 W 990 Industrial Rd.
Barrington, IL 60010
(312) 381-7441
(312) 381-7453

alcohol distilling plants with 200 gallons per day output

Chemapec Inc.
230 Crossways Park Dr.
Woodbury, NY 11797
(516) 364-2100

design, process, procurement, construction, management; plant size: 5 million gallons per year

Jerry Childress
P. O. Box 47
Baring, MO 63531
(816) 397-3960

used stainless steel ethanol columns

Clary Corporation
917 Parkway Dr.
Grand Prairie, TX 75051
(214) 647-4156

manufactures 70 to 200 gallons per day distilleries

Conklin Co.
4660 W. 77th Street
Minneapolis, MN 55410
(612) 831-4044

sells units with 50,000 to 100,000 gallons per year capacity

Conrad Industries
Box 130
Bonaparte, IA 52620
(319) 592-3131

manufactures farm stills of various sizes, 200,000 to 2 million gallons capacity

Coulter Copper and Brass
140 The East Mall
Toronto, Ontario
Canada M8Z 2M2
(416) 239-2771

still design, components, engineering

Davy McKee Corp.
10 South Riverside Plaza
Chicago, IL 60606
(312) 454-3810

engineering, procurement, design consulting and technology; plant size: 5 million to 20 million gallons per year

Day & Zimmerman, Inc.
1818 Market Street
Philadelphia, PA 19103
(215) 299-8193

complete plant construction and services for plants of all sizes

Domestic Technology Institute
P. O. Box 2043
Evergreen, CO 80239
(303) 988-3054

blueprints, engineering specifications, training, plant size: 45,000 gallons per year

Double "A" Products
Division of Ad-Art Inc.
P. O. Box 1107
Albert Lea, MN 56007
(507) 373-1458

farm-size distillation plants

Easy Engineering
3353 Larimer
Denver, CO 80205
(303) 893-8936

alcohol plants with 20-25 gallons per hour capacities

Elmwood Energy, Inc.
Box 321
Plainfield, IL 60544
(815) 436-7463

stills with capacities of 150 gallons per 24 hours

Energy Independence Corp.
Box 56
Montrose, MN 54363
(612) 367-3131

units with capacities of 20 gallons per hour

Energy Restoration Inc.
1201 J St.
Century House Suite 101
Lincoln, NE 68508
(402) 475-9237

plants with capacities of 60 and 120 gallons per hour

Ernest Gage Co.
250 S. Livingston Ave.
Livingston, NJ 07039
(201) 992-1400

components

Ethanol International, Inc.
Marketing Office, Suite 200
5350 So. Denver Tech Center Pkwy.
Englewood, CO 80111
(303) 741-2810

small to large-scale alcohol plants; complete services

Farm Fuels, Inc.
Energy Center One
717-17th Street
Suite 1535
Denver, CO 80202

still distributors

Flexi Liners
5940 Reeds Rd.
Mission, KS 65775
(913) 372-4331

fermentation tank liners

Fort Wayne Dairy Equipment Co.
Box 269
Fort Wayne, IN 46801
(219) 424-5517

components

Fuel and Power, Inc.
904 Oakdale
Court, MI 49423
(616) 399-2436

units that will produce 10 gallons per hour

Glitsch, Inc.
P. O. Box 226/227
Dallas, TX 75266
(214) 631-3841

skid-mounted distillation units with 1,000 gallons per day output; larger units also will be available

Homak, Inc.
Box 3242
Abilene, TX 79604
(915) 673-4015

engineering and manufacturing of extruders of several sizes

In-Sol Energy
P. O. Box 971
Taylor, TX
(512) 352-5513

sales agent for Gasohol, Inc., a subsidiary of Coulter, Copper & Brass, Toronto, Ontario; sells stills of all sizes

K A M Products Corp.
98-21 97th Ave.
Ozone Park, NY 11416
(212) 845-4600

manufacturers of heating, cooling coils

KBK Industries
E. Highway 96
Rush Center, KS 67575

fiberglass fermentation tanks

Kargard Industries
Marinette, WI 54143
(715) 735-9311

stainless steel tanks

L & A Engineering and Equipment
4124 South Soderquist Rd.
Turlock, CA 95380
(209) 632-3191

designs and constructs commercial stills of all sizes

McEver Engineering Inc.
6363 Richmond Ave.
Houston, TX
(713) 780-3465

detailed mechanical and electrical design construction

Meters and Gages
Box 400
Marshalltown, IA 50158
(515) 752-9299

components

Middle State Mfg. Co.
16th Ave., Box 788
Columbus, NE 68601
(402) 564-1411

component manufacturer

Mosier Construction
511 N. Cedar
P. O. Box 381
Owatonna, MN 55060
(507) 451-5254

builds units with capacities of 20 to 60 gallons per hour

OMTEC
Omaha Tank and Equipment
13706 Giles Rd.
Omaha, NE 68138
(402) 896-1800

tanks, columns and distillation equipment

Osage Plains, Inc.
P. O. Box 287
Butler, MO 64730
(816) 679-3842

developing plant prototypes

PEDCo International, Inc.
11499
Chester Rd.
Cincinnati, OH 45246
(513) 782-4500

designs, engineers and supervises construction of plants of many sizes

Process Group:
   DeLaval Deparator Group
139 Gaither Drive
Mt. Laurel, NJ 08054

engineering, design, technology

Jim Pufahl
RR #2, Box 99
Milbank, SD 57252

12-inch diameter distillation columns up to 16 inches high made from mild steel; stainless steel columns

R.B. Industries
Box 82
Riverdale, MI 48877
(517) 833-7584

developing an alcohol extraction process that does not require distillation

Raphael Katzen Associates
1050 Delta Ave.
Cincinnati, OH 45208

consulting — technical and economic, engineering and feasibility studies, process and equipment design.

Real Fuel Corp.
4320 Abbott Avenue South
Minneapolis, MN 55410
(612) 929-8272
(612) 926-5669

distributor and on-site fabricator of alcohol fuel operations, using methane where possible; plants with 500,000 and 1 million gallons per year capacities

Rochelle Development Inc.
Box 356
Rochelle, IL 61068
(815) 562-7372

component manufacturers; "turn-key" alcohol plants

Rush Tower Alcohol Fuels Co.
Weaver Road
RR1, Box 282A
Festus, MO 63028
(314) 937-2831

sells units produced by Easy Engineering and A.M.F. Padovan of Italy; units produce 25 to 500 gallons per hour

Silver Engineering Works Inc.
3309 Blake St.
Denver, CO
(303) 623-0211

component manufacturers; engineering, procurement, construction, design, consulting, and technology; plant size: 100,000 to 20 million gallons per year

Sludge Express Co.
Sheldon, IA 51201
(712) 324-3305

component manufacturers

Solar Energy Innovations, Inc.
54-45 44th Street
Maspeth, NY 11378
(212) 361-9038

distillation system that produces 30 to 40 gallons per day

Solstice Designs, Inc.
P. O. Box 2043
Evergreen, CO 80439

construction blueprints, information packages

Stilco Engineering Corp.
Rt. 3, Box 177
Comanche, TX 76552

engineering, construction

Still Company
P. O. Box 9750
115 E. 9th St.
Panama City Beach, FL 32407

stills, equipment, consulting

Stone & Webster Engineering Corp.
2600 River Road, Suite 202
Des Plaines, IL 60018
(312) 827-6791 (Chicago district
    office)

full range of engineering services including site studies and construction; plant size: 55,000 to 20 million gallons per year

Stone & Webster Engineering Corp.
One Penn Plaza
250 West 34th St.
New York, NY 10001 (New York
    office)

Sun Power, Inc.
8301 Braniff
Houston, TX 77061
(713) 645-7173

four models of refineries for making alcohol; fermentation tanks

SUN-STOR Corp.
1583 E. 580 S.
Pleasant Grove, UT 84062
(801) 875-5625

alcohol fuels, solar engineering, resource recovery

3T Engineering Inc.
Box 80
Arenzville, IL 62611
(217) 997-5921

component manufacturers

Tri-Star Corp.
RR 3
Illini Street
Vandalia, IL 62471
(618) 283-1666

*easy-to-operate small plants with 35
gallons per day capacity*

Union Development Company
Suite 409, Exchange Center II
4606 S. Garnett
Tulsa, OK 74145
(918) 627-0010

*design and plant construction services
for plants with 2,000 to 20,000 gallons
daily capacity; distillation facilities
with capacities of about 500,000 to 10
million gallons per year*

United International, Inc.
P. O. Box 11
Buena Vista, GA 31803
(912) 649-7449

consultants, designers, builders

VARA International, Inc.
1201 19th Place
Vero Beach, FL 32960
(305) 567-1320

*alcohol fermentation and distillation
systems with 500,000 to 20 million
gallons annual capacity*

Vendome Copper and Brass
153 North Shelby St.
Box 1118
Louisville, KY 40202
(502) 587-1930

*a major component manufacturer that
has produced stills, columns,
condensers and other distillation
equipment with capacities of more
than 100,000 gallons per day*

Vogelbusch Division
Bohler Bros. of America, Inc.
1625 West Belt North
Houston, TX 77043
(713) 465-8376

*designed and built large-scale plants in
Brazil, United States and Europe; a
leading company producing plants
with 500,000 to 100 million gallons
annual capacity*

Vulcan Cincinnati, Inc.
2900 Vernon Place
Cincinnati, OH 45219
(513) 281-2800

designs, constructs large stills

W.A. Bell
P. O. Box 105
Florence, SC 29503

component manufacturer

World Energy Co.
RR 5
Box 251
Carthage, MO 64836
(417) 358-8174

sells units capable of producing up to 75 gallons of alcohol per day

World-Wide Construction
  Services, Inc.
P. O. Box 8126
Wichita, KS 67208
(316) 942-0101

still design, engineering, equipment procurement and still fabrication

Zeithamer Enterprises Inc.
Route 2, Box 63
Alexandria, MN 56308
(612) 763-7392
(612) 762-1798

alcohol fuel equipment for farm-size stills; one of the pioneers in home-built stills; has on-site plant with 50,000 gallons annual capacity

Miscellaneous Material and Equipment Suppliers for Farm and Large Still Users

APV Company, Inc.
395 Fillmore Ave.
Tonawanda, NY 14150
(716) 692-3000

designs and fabricates many types of distillation equipment

Ag-Bag Corp.
P. O. Box 218
Astoria, OR 97103
(503) 325-2488

gasohol product storage equipment

Alcohol Plant Supply Co.
P. O. Box 248
Sherwood, OR 97140
(503) 244-3230

alcohol fuel and methane plants and service

Alcohol Technology, Inc.
P. O. Box 1489
Rockdale, TX 76567
(512) 446-6777

alcohol fuel plants with a capacity of 30 to 50 gallons per hour

Alternative Energy
P. O. Box 353
Colby, KS 67701
(913) 462-7171

design, construction, consulting and training; plant size: 50,000 to 150,000 gallons per year

Arbor Sales Co., Ltd.
P. O. Box 6
Des Moines, IA 50301

distillation plates and columns alcohol stills

Arthur H. Thomas Co.
Vice St. at 3rd
P. O. Box 779
Philadelphia, PA 19105
(212) 574-4500

proof hydrometers

BIOCON U.S., INC.
261 Midland Ave.
Lexington, KY 40507
(606) 254-0517

yeasts, enzymes, foaming controls

Bryan Steam Corp.
P. O. Box 27
Peru, IN 46970
(317) 473-6651

boilers engineered especially for ethanol production

Carolina Biological Supply Co.
Burlington, NC 27215

proof hydrometers, laboratory equipment

Conbraco Industries, Inc.
P. O. Box 125
Pageland, SC 29728
(803) 672-6161

safety valves, valves, water gauges

DADY
Universal Foods Corp.
P. O. Box 737
Milwaukee, WI 53201
(414) 271-6755

Red Star Distiller's Active Dry Yeast (DADY)

Flexi Liners
5940 Reeds Rd.
Mission, KS 66202
(913) 265-1234

fermentation tank liners

GB Fermentation Industries, Inc.
One North Broadway
Des Plaines, IL 60016
(312) 827-9700

yeasts, enzymes

Gallard-Schlesinger Chemical
   Mfg. Corp.
584 Mineola Ave.
Carle Place, NY 11514
(516) 333-5600

pH test papers

Grove Engineering
1714 Gervais Ave.
North St. Paul, MN 55109
(612) 777-8545

Hi Bar   *products for nitrogen fixation,*
P. O. Box 2075   *biocatalyst*
Oroville, CA 95965

InstaPro   *extruders designed to cook and*
10301 Dennis Drive   *gelatinize the starch in corn for*
Des Moines, IA 50322   *alcohol production; external heat*
(515) 276-4524   *sources not needed*

Jacobson Machine Works, Inc.   *hammer mills, grinders and processing*
2445 Nevada Ave. North   *equipment*
Minneapolis, MN 55427

M & W Gear   *water/alcohol injection in diesel*
Route 47 South   *engines*
Gibson City, IL 60936
(217) 784-4261

Manning & Lewis Engineering Co.   *shell and tube heat exchangers,*
679 Rahway Ave.   *condensers*
Union, NJ 07083
(201) 687-2400

Meters and Gages   *components*
Box 400
Marshalltown, IA
(515) 752-9299

Middle State Manufacturing, Inc.
P. O. Box 788
Columbus, NE 68601
(402) 564-1411

Miles Laboratories, Inc.   *enzymes*
Enzyme Products Division
1127 Myrtle St.
P. O. Box 932
Elkhart, IN 46515
(219) 262-7176

Novo Laboratories Inc.   *enzymes*
59 Danbury Road
Wilton, CT 06897

The Protectoseal Company
225 West Foster Avenue
Bensenville, IL 60106
(312) 595-0800

storage tank fittings and safety devices for alcohol and gasohol storage tanks

Semplex of U.S.A.
P. O. Box 12276
Minneapolis, MN 55416
(612) 522-0500

hydrometers, wine yeasts, and so on

Tomco Equipment
1862 S. Duth
Hidden Hills Parkway
Stone Mountain, GA 30088
(404) 466-2235

co-producers and manufacturers of storage equipment

United International Research, Inc.
230 Marcus Boulevard
Hauppauge, L.I., NY 11787
(516) 273-0900

sells Hydrelate, *a special stabilizer that offsets problems caused by residual water in alcohol fuels*

Universal Foods Corp.
433 East Michigan St.
Milwaukee, WI 53201

distillers, yeasts

Suggested Checklist for 55-Gallon Still (Using BIOCON Enzymes and Yeasts)

MONTH _____

DATE_____

TIME OF START _____

COOKING

Flue secure	_____
Materials on hand	_____
Doors and windows open	_____
Snow or water on hand	_____
Fire extinguisher	_____
55-gallon drum clean	_____
Clean-out door tight	_____
Gate valve shut	_____
Top hatch plate in place	_____
Add 20 to 25 gallons of water	_____
Add 56 pounds corn meal (one bushel)	_____
Add 3 teaspoons *Canalfa*	_____
Light fire	_____(Time)

Insulation on   _____

Increase to 200° F.   _____

Reach 200° F.   _____(Time)

Hold 15 minutes at 200° to 212° F.   _____

Insulation off   _____

Retard flame   _____

Drain mash   _____

Scrape mash from drum   _____

Cool to 150° to 170° F.   _____

Reach 170° F.   _____(Time)

Add 6 teaspoons *Canalfa*   _____(Time)

Hold 30 minutes at 150° to 170° F.   _____

Cool to 120° F.; reach 120° F.   _____(Time)

Add 6 teaspoons *Gasolase*   _____(Time)

Sugar test (strain mash first)   _____(Percent)

FERMENTATION

Cool to 100° F.   _____

Reach 100° F.   _____(Time)

Add 12 teaspoons of yeast   _____

Ferment three to six days (optimum temp. 86° F.)   _____

Stir, check temperatures   _____

Gases stop   _____(Date, time)

Sugar test   _____(Percent)

DISTILLATION

Flue secure   _____

Materials and equipment on hand   _____

Doors, window open   _____

Water on hand   _____

Fire extinguisher   _____

55-gallon drum clean   _____

Clean-out door tight   _____

Top hatch plate in place   _____

Check pressure release valve   _____

Check column and copper lines   _____

Separate solids and liquid   _____

Pour liquid into drum (no more than
    ⅔ full)   _____(Quantity)

Fill cooling tub with water   _____

Light fire   _____(Time)

Tighten up hatch   _____

Place plastic container near "tail"   _____

Insulation on   _____

Check temperature/pressure gauge   _____

Heat to 190° to 200° F.   _____

Reach 190° to 200° F.   _____(Time)

First shots   _____(Time, proof)

Alcohol proof 30 minutes later   _____

Proof declines to 100 or below   _____(Time)

Run complete   _____

APPENDIX E

Resource People and Organizations

American Agriculture Foundation, Inc.
Box 57, Springfield, CO 81073
(303) 523-6223

American Agriculture Movement
308 Second Street, SE
Washington, D.C. 20515
(202) 544-5750

The Bio-Energy Council
1625 Eye Street, NW
Suite 825A
Washington, D.C. 20006
(202) 833-5656

National Farmers Union
Brewers Grain Institute
1750 K Street, NW
Washington, D.C. 20006

Corn Development Commission
Route 2
Holdrege, NE 68949

Distillers Feed Research Council
1435 Enquirer Building
Cincinnati, OH 45202
(513) 621-5985

Gasohol USA
10008 East 60th Terrace
Kansas City, MO 64133
(816) 737-0064

Iowa Corn Promotion Board
200 West Towers
1200-35th St.
West Des Moines, IA 50265
(515) 225-9242

Mid-America Solar Energy Complex
8140-26th Ave., South
Bloomington, MN 55420
(612) 854-0400

National Center for Appropriate Technology
P. O. Box 3838
Butte, MT 59701
(406) 494-4572

National Farmers Organization
Surprise, NE 80251

National Farmers Union
Denver, CO 80251

National Gasohol Commission, Inc.
521 South 14th Street, Suite 5
Lincoln, NE 68508
(402) 475-8044 or 8055

Small Farm Energy Project
P. O. Box 736
Hartington, NE 68739
(402) 254-6893

The Grange
Route 1, Box 154
Waterloo, NE 68069
(402) 359-5605

International Biomass Institute
1522 K Street, NW
Suite 600
Washington, D.C. 20005
(202) 783-1133

Solar Energy Research Institute
1617 Cole Boulevard
Golden, CO 80401
(303) 231-1207

The Wheat Growers
Route #1, Box 27
Hemingford, NE 69438
(308) 487-3794

Colleges

Brevard Comm. College
1519 Clearlake Road
Cocoa, FL 32922
Contact: Maxwell King
(305) 632-1111

Cecil Comm. College
1000 North East Road
North East, MD 21901
Contact: Robert Gell
(301) 287-6060

Clark University
450 Main
Worchester, MA 01610
Contact: Harry C. Allen
(617) 793-7711

College of Siskiyous
800 College Avenue
Weed, CA 96094
Contact: Gary Peterson
(916) 938-4463

College of Southern Idaho
315 Falls Avenue West
Twin Falls, ID 83301
Contact: James Taylor
(208) 733-9554

College of the Virgin Islands
St. Thomas, VI 00801
Contact: Michael Canoy

Delaware Tech. & Comm. College
1832 N. Dupont Parkway
Dover, DE 19901
Contact: Rich Morchese
(302) 678-5416

Des Moines ATVI
2006 South Ankeny Blvd.
Ankeny, IA 50021
Contact: Richard Byer
(515) 964-6228

Eastern Iowa Comm. College
2804 Eastern Avenue
Davenport, IA 52803
Contact: Robert Illingsworth
(319) 242-6841

Eastern Oregon State College
8th & K Streets
LaGrande, OR 97850
Contact: Terry Edvalson
(503) 963-2171

Eastern Wyoming College
3200 West C
Torrington, WY 82240
Contact: Charles Rogers
(307) 532-7111

Iowa Central Comm. College
330 Avenue M
Fort Dodge, IA 50501
Contact: Edwin Barbour
(515) 576-3103

Kankakee Comm. College
Box 888
Kankakee, IL 60901
Contact: M.E. Marlin
(815) 933-0345

Lake Land Comm. College
South Route 45
Matoon, IL 61938
Contact: Robert D. Webb
(217) 235-3131

Lamar Comm. College
2401 South Main
Lamar, CO 81052
Contact: Bill Henderson
(303) 336-2248

Lehigh County Comm. College
2370 Main Street
Schnecksville, PA 18078
Contact: Robert Walker
(215) 799-1141

Lincoln Land Comm. College
Springfield, IL 62708
Contact: Robert Poorman
(217) 786-2200

Mid-South Energy Project
Mississippi County Comm. College
Box 1109
Blytheville, AR 72315
Contact: Harry Smith
(501) 762-1020

Modesto Jr. College
Modesto, CA 95350
Contact: Ron Alver
(209) 526-2000

Mott Comm. College
1401 East Court Street
Flint, MI 48503
Contact: Charles Roche
(313) 762-0237

Navajo Comm. College
Box 580
Shiprock, NM 87420
Contact: Raymond Housh
(505) 368-5291

Navarro Jr. College
Box 1170
Corsicana, TX 75110
Contact: Darrell Raines
(214) 874-6501

Nicholls State University
Thibodaux, LA 70301
Contact: William Flowers
(504) 446-8111

North Dakota State School
of Science
Wahpeton, ND 58075
Contact: Claire T. Blikre
(701) 671-2221

NW Mississippi Junior College
Highway 51
North Senatobia, MS 38668
Contact: William Oakley
(601) 562-5262

Oglala Sioux Comm. College
Box 439
Pine Ridge, SD 57700
Contact: Roberta Barbalace
(606) 867-5110

Onondaga County Comm. College
Syracuse, NY 13215
Contact: Andreas Paloumpis
(315) 469-7741

Paducah Comm. College
Box 1380
Paducah, KY 42001
Contact: Donald Clemons
(502) 442-6131

Panhandle State University
Box 430
Goodwell, OK 73939
Contact: Gene Reeves
(405) 742-2121

Pitt Comm. College
Box Drawer 7007
Greenville, NC 27834
Contact: William Fulford
(919) 756-3130

South Dakota State Univ.
Brookings, SD 57007
Contact: Paul Middaugh
(605) 688-4111

South East Comm. College
Milford, NE 68405
Contact: Dean Roll
(402) 761-2131

Southwest State University
Marshall, MN 56258
Contact: Richard Spencer
(800) 533-5333

Springfield Tech. & Comm. College
One Armory Square
Springfield, MA 01105
Contact: Robert Geidz
(413) 781-6470

State Fairground Comm. College
Sedalia, MO 65301
Contact: Marvin Fielding
(816) 826-7100, Ext. 60

Talladega College
627 West Battle Street
Talladega, AL 35160
Contact: Richard A. Morrison
(205) 362-8800

Texas Tech. University
Lubbock, TX 79409
Contact: Steven R. Beck
(806) 742-2121

University of Vermont
Burlington, VT 05405
Contact: Robert B. Lawson
(802) 656-2990

Vincennes University
Vincennes, IN 47591
Contact: Daryle Riegle
(812) 882-3350

Washington State University
Box 708
Chehalis, WA 98532
Contact: Larry Gueck
(206) 748-9121, Ext. 212

Miscellaneous Sources

Alcohol Fuel Directory Service
P. O. Box 2213
Canoga Park, CA 91303

addresses and phone numbers for more than 650 sources; manufacturers, financial aid sources, publications

Department of Energy
Office of Public Affairs
8G031 Forrestal Building
Washington, D.C. 20585
(202) 252-5568

G.V. Olsen Associates
170 Broadway
New York, NY 10038
(212) 866-5034

MAR-CAM Industries, Inc.
758 N. Easton Road
Glenside, PA 19038
(215) 887-9070

specializes in marketing, distribution and sale of agricultural alcohol and gasohol

National Technical Information
   Service
U. S. Department of Commerce
5285 Port Royal Road
Springfield, VA 22161

Sources of Public Financing

Organization/ Additional Information	Program/ Eligible Applicants	Type of Assistance
U.S. Department of Agriculture Science & Education Adm. Washington, D.C. 20250 (202)447-6050	Alcohol & Industrial Hydrocarbons (Sect. 1419 of Food and Agricultural Act of 1977, P.L. 95-113) for colleges and universities having a demonstrable capacity in food and agricultural research	Grants of 2-3 years duration for research
U.S. Department of Agriculture Science & Education Admin. Washington, D.C. 20250 (202)447-6050	Energy Research (Sect. 1414 of Food and Agricultural Act of 1977, P.L. 95-113) for colleges and universities having a demonstrable capacity in food and agricultural research	Grants of 2-3 years duration for research
U.S. Department of Agriculture/Office of Energy Rm. 226-E, Administration Washington, D.C. 20250 (202)447-2455	General advice; no restrictions	General advice on USDA program availability
U.S. Department of Agriculture/Farmers Home Administration Director of B&I Loans Washington, D.C. 20250 (202)447-7595 (202)447-5243	Business & Industrial (B&I) for cooperatives, private investors in town of less than 50,000	Loan guarantees
U.S. Department of Agriculture/Farmers Home Admin. Director of Farm & Family Programs—USDA/FmHA Washington, D.C. 20250 (202)447-4671 (202)447-4597	Operating and farm ownership loans for farmers, farmer cooperatives	Direct loans at cost of borrowing, loan guarantees

Eligible Activities	Purpose of Project	Limits of Project
Research on the evaluation, handling treatment and conversion of biomass resources for manufacture of ethyl alcohol	To develop improved processes for production of alcohol from biomass	$100,000 per grant of 2-3 years duration
Research on fermentation and related processes for production of alcohol, other than ethanol; and hydrocarbons	To develop improved methods of production and blending, marketing and utilization of products	$100,000 per grant of 2-3 years duration
Biomass production for alcohol fuels; conversion and use of alcohol	Serves as information clearinghouse and provides for coordinated USDA	None
Fixed costs, operating capital	Creation of jobs, economic growth in communities under 50,000 population	$25,000,000 per project maximum; priority on small and intermediate scale of $1,000,000 or less
Fixed assets, working capital	Improvement of farm income	$200,000 direct loan; $300,000 loan guarantee

Organization/ Additional Information	Program/ Eligible Applicants	Type of Assistance
U.S. Department of Agriculture/Farmers Home Admin. Director of Rural Development Programs — USDA Washington, D.C. 20250 (202)447-3213	Community facilities for private, non-profit, public entities	Loans at 5%
Housing and Urban Development (HUD) Director of UDAG/HUD Washington, D.C. 20410 (202)472-3947	Urban Development Action Grant for distressed cities and urban counties	Grant to city to be used for public improvements or loans to developer
Technology Assistance Division, Small Business Admin. Washington, D.C. 20416 (202)653-6586	Small Business Energy Loan Act, P.L. 95-313 for small business, including farmers and cooperatives for solar and energy conservation technologies	Loans and loan guarantees
Department of Commerce Economic Development Administration EDA Office of Public Affairs Washington, D.C. (202)377-5113	Public works and development facilities for states, local governments, Indian tribes, non-profit organizations	Grants for 50-80% of a total project cost depending on need
Economic Development Administration EDA Office of Public Affairs Washington, D.C. (202)377-5113	Business development assistance for business enterprises including cooperatives	Direct loans up to 65%; loan guarantees up to 90%
Energy Programs Office of Comm. Action Community Services Adm. 1200 19th St. N.W. Washington, D.C. 20506 (202)632-6503	Grants (limited) — technical assistance for rural and small farm energy grantees	Construction and operation of demonstration plants serving energy needs of rural low income residents, provision of technical assistance to other communities in small scale alcohol production

Eligible Activities	Purpose of Project	Limits of Project
Construction loans, working capital	Improvement of the levels of public services and economic growth	Same as B&I ($25,000,000 project maximum); priority on small and intermediate scale of $1,000,000 or less
Fixed assets, related expenses	Stimulate employment and tax base in distressed cities	None
Working capital, research and supplies, plant construction, materials, development, manufacturing equipment for alcohol fuels purposes	Promote small businesses in alcohol production related activities	Direct loans of less than $350,000, loan guarantees of less than $500,000. No more than 30% for R&D, no more than 35% for working capital
Construction and equipment of alcohol fuel plants, priorities on small scale plants (less than one million gallons per year)	To stabilize or stimulate local economy, agricultural area emphasis	Generally $300,000 per project, must be in EDA designated redevelopment area
Fixed asset and/or working capital for production plants or auxiliary facilities to such plants	Help job situation, increase income, increase crop markets, increase supply of transportation fuel	Generally for $500,000 minimum size. Plants must be in eligible areas. This program normally would not be appropriate for individual farmers
To develop and disseminate efficient technologies for small scale fuel alcohol production	Grants go only to 5 currently funded CSA projects. Phase II technical assistance available to other eligible organizations. Phase III	

Organization/ Additional Information	Program/ Eligible Applicants	Type of Assistance
Solar Energy Research Institute 1617 Cole Boulevard Golden, CO 80401 (303)231-1415	Biomass Energy Systems Program for individuals, farmers, businesses, institutions (no restrictions)	Technical assistance, competitive awards
Department of Energy DOE Regional Offices	Small Scale Technology Program for individuals and small institutions	Grants
Chief, Alternative Fuels Utilization Program Department of Energy 20 Mass. Ave., N.W. Washington, D.C. 20528	Alternative Fuels Utilization Program for individuals, farmers, businesses, institutions (no restrictions)	Competitive awards
Chief, Community Technology Systems Branch Department of Energy 20 Mass. Ave., N.W. Washington, D.C. 20528	Urban waste programs for individuals, businesses, institutions, communities (no restrictions)	Competitive awards— loan guarantees are under consideration

DOE Regional Office Contacts

Use the address for the region where your state is located.

States by Region
1. VT, MA, CT, RI, NH, ME
2. NY, NJ
3. VA, WV, MD, DE, PA
4. KY, TN, NC, SC, GA, AL, MS, FL
5. MN, WI, IL, IN, OH, MI
6. NM, TX, OK, AR, LA
7. NE, KS, MO, IA
8. CO, UT, WY, MT, ND, SD
9. HA, CA, NV, AZ
10. AK, WA, OR, ID

DOE Region 1 Office
Bob Chase*
(617)223-3106
150 Causeway Street
Boston, MA 02114

DOE Region 2 Office
Jane Delgado*
(212)264-0520
26 Federal Plaza
New York, NY 10007

DOE Region 3 Office
Tony Pontello*
(215)597-3607
1421 Cherry Street
Philadelphia, PA 19102

*Contact for Small Scale Grants Program

Eligible Activities	Purpose of Project	Limits of Project
Conversion of biomass to alcohol fuels	Research and development for on-farm systems, advanced energy crops, collection and harvesting improvements, and advanced conversion technologies	None
Small scale renewable energy sources	Develop innovative small-scale renewable energy technologies	$50,000 per project over two years
Research and development — also testing of alternative fuels	Develop and test alternative fuels including alcohols in diesel and internal combustion engines	None
Conversion of urban and municipal waste products to energy	Conduct research and development and demonstrate techniques converting municipal waste to gases and liquids energy	None

DOE Region 4 Office
Micky Feltus*
(404)881-2386
1655 Peachtree St. NE
Atlanta, GA 30309

DOE Region 5 Office
John Purcell*
(312)972-2067
Roberta Dalton*
(312)972-2383
175 W. Jackson Blvd.
Chicago, IL 60604

DOE Region 6 Office
Chuck Royston*
(214)767-7777
P. O. Box 35228
Dallas, TX 75235

DOE Region 7 Office
Jack Stacy*
(816)374-3815
324 E. 11th St.
Kansas City, MO 64106

DOE Region 8 Office
Tom Stroud*
(303)234-2165
P.O. Box 26247,
Belmar Br.
Lakewood, CO 80226

DOE Region 9 Office
Meg Schachter*
(415)556-1465
111 Pine Street
San Francisco, CA 94111

DOE Region 10 Office
Frank Brown*
(206)442-1746
915 Second Ave.
Seattle, WA 98174

National DOE Hotline
(answers questions on
alcohol financing
and construction)
(800)535-2840

Suggested Additional Reading

A Learning Guide for Alcohol Fuel Production. Colby, Kansas: Colby Community College, 1255 South Range, 67701. 356 pages, tables, diagrams, photos, $45. For the serious alcohol producer, this book is a worthwhile investment. Thorough, but graphics are of inconsistent quality; written with the help of several experts, including Dr. Paul Middaugh of South Dakota State University. Complete discussion of alcohol production and plant construction.

Crombie, Lance. *Making Alcohol Fuel, Recipe and Procedure.* Minneapolis, Minnesota: Rutan Publishing, P. O. Box 3585, 55403. 115 pages. This is another basic beginner's book on making alcohol fuel, including thorough discussion of fermentation. Equipment and plant designs are discussed, but no plans. One of the best and earliest books in this field.

Fuel From Farms: A Guide to Small-Scale Ethanol Production. Golden, Colorado: Solar Research Institute, 1617 Cole Boulevard 80401. 75 pages, plus seven appendices. Available from two other sources: The National Technical Information Service, U.S. Department of Commerce, 5285 Port Royal Road, Springfield, Virginia, 22161 and The Superintendent of Government Documents, U.S. Government Printing Office, Washington, D.C. 20402. An excellent, comprehensive discussion of making alcohol on the farm. Discussion of plant design and operation. Large farm units considered, no plans. Thorough consideration of basic production processes, alcohol market assessment, financial requirements, planning, by-products and business planning. A solid, serious book.

Gibat, Kathleen and Norman. *The Lore of Still Building.* Fostoria, Ohio: Popular Topics Press, Box 1004, 44830. 128 pages, small paperback, $5. Written in an informal, easy-going manner, this book is intended for those who want to make alcoholic *beverages.* For the beginning alcohol fuel maker, it's also worthwhile reading as explanations are clear, concise. Includes helpful tables.

Manual for the Home and Farm Production of Alcohol Fuels. Los Banos, California: J.A. Diaz Publishing Co., Box 709, 93635. 127 pages. Includes 26 pages on basic fuel theory. Covers all alcohol production steps. Information on still designs is sketchy but, on the whole, the book is a helpful addition to the beginning fuelmaker's library. No plans.

Willkie, Herman F. and Kolachov, Paul J. *Food for Thought*. Indianapolis, Indiana: Indiana Farm Bureau, Inc. 209 pages, small paperback, $2. First printed in 1942, this book is an excellent source of information about the characteristics of ethanol, distillation experiments and test results and basic fermentation and distillation processes. It was not intended as a practical, how-to guide, but the book is highly regarded by many people interested in distillation.

Historical Reading

Carr, Jess. *The Second Oldest Profession: An Informal History of Moonshining in America*. Englewood Cliffs, New Jersey: Prentice-Hall. A native of southwest Virginia, Carr has lived and worked among the mountain people most of his life. In this book, he has reconstructed an excellent history of moonshining. More detailed, but not quite as colorful as Dabney's *Mountain Spirits*. Both books make delightful reading. Both authors have spoken with many, many moonshiners and others familiar with the subject.

Dabney, Joseph Earl. *The Corn Whiskey Recipe Book*. Atlanta, Georgia: Sassafras Press, 3966 St. Clair Court, 30319. 77 pages, $2.95 plus 20 cents postage. For the person who wants to learn more about illicit alcohol making in the South, this is a good buy. There's how-to information on moonshine still-making, profiles of moonshiners, "trippers" and federal agents.

Dabney, Joseph Earl. *Mountain Spirits: A Chronicle of Corn Whiskey from King James Ulster Plantation to America's Appalachians and the Moonshine Life*. Lakemont, Georgia: Copple House Books. 242 pages. Here's what *Time* magazine said about this book: "From dusty books and first hand interviews with oldtimers, with many facts and much affection, Joseph Dabney has put together a splendid and sometimes hilarious history."

Other Helpful Books

Brown, Michael H. *Brown's Alcohol Motor Fuel Cookbook*. Cornville, Arizona: Desert Publications. 140 pages, $9.95. About 50 pages are devoted to carburetor adjustment and changes needed to operate a car on alcohol. There's also discussion of fermentation and moonshine-type distillation. Plans are not included.

Daniels, George. *Home Guide to Plumbing, Heating and Air Conditioning*. New York, New York: Popular Science/Outdoor Life Book Division, Times Mirror Magazines, Inc. 186 pages. Helpful information on the tools, techniques, equipment for plumbing, useful for those who want to fabricate stills from readily available materials.

Distillers Feeds. Cincinnati, Ohio: Distillers Feed Research Council, 1435 Enquirer Building, 45202. 116 pages. A comprehensive, scholarly discussion of distillers feeds, the high-protein by-product of distillation. Result from test using distillers feeds for cattle, sheep, swine and other animals.

Feed Formulation.   Cincinnati, Ohio: Distillers Feed Research Council, 1435 Enquirer Building, 45202. A brief discussion of distillers feed in poultry, cattle, swine and other rations.

Mother's Alcohol Fuel Seminar.   The Mother Earth News, Inc., P. O. Box 70, Hendersonville, N.N. 29739, 54 pages, $25. A helpful, but high-priced, looseleaf book that covers all the basics of making alcohol fuel; includes reprints of several articles that first appeared in the *Mother Earth News.*

Peterson's How to Tune Your Car.   Los Angeles, California: Peterson's Publishing Co., 8490 Sunset Boulevard, 90069. 256 pages, $5.95. A good reference book for those who want to adjust their carburetors, ignition systems. Well illustrated.

Shelton, Jay. *Wood Heat Safety.*   Charlotte, Vermont: Garden Way Publishing. 165 pages, $8.95. A definitive discussion of all aspects of heating with wood safely. Discussion of flue requirements, clearances and boiler pressures is appropriate for still builders.

Wik, Ole. *Wood Stoves: How to Make and Use Them.*   Anchorage, Alaska: Northwest Publishing Co. 194 pages. A good guide for making stoves from barrels and other used materials. Helpful for those who want to fabricate stills from 55-gallon drums.

Bulletins, Magazines

ADM Gasohol News
ADM Corn Sweeteners
1350 Waconia Ave. SW
Cedar Rapids, IA 52406

Gasohol U.S.A.
P. O. Box 9547
Kansas City, MO 64133

Latin American Energy Report
Business Publishers, Inc.
P. O. Box 1067
Silver Springs, MD 20910

The Mother Earth News
P. O. Box 70
Hendersonville, NC 28739

The Small Farm Energy Project Newsletter
Small Farm Energy Project
Center For Rural Affairs
P. O. Box 736
Hartington, NE 68739

Glossary

ALCOHOL: the family name of a group of organic chemical compounds composed of carbon, hydrogen, and oxygen; a series of molecules that vary in chain length and are composed of hydrocarbon plus a hydroxyl group, $CH_3\text{-}(CH_2)n\text{-}OH$; includes methanol, ethanol, isopropyl alcohol, and others.

ALKALI: soluble mineral salt of a low density, low melting point, highly reactive metal; characteristically "basic" in nature.

ALPHA-AMYLASE-AMYLASE: enzyme which converts starch into sugars.

ANAEROBIC DIGESTION: without air; a type of bacterial degradation of organic matter that occurs only in the absence of air (oxygen).

ANHYDROUS: a compound that does not contain water either absorbed on its surface or as water of crystallization.

ATMOSPHERIC PRESSURE: pressure of the air (and atmosphere surrounding us) which changes from day to day; it is equal to 14.7 psia.

AZEOTROPE: the chemical term for two liquids that, at a certain concentration, boil at the same temperature; alcohol and water cannot be separated further than 194.4 proof because at this concentration, alcohol and water form an azeotrope and vaporize together.

AZEOTROPIC DISTILLATION: distillation in which a substance is added to the mixture to be separated in order to form an azeotropic mixture with one or more of the components of the original mixture; the azeotrope formed will have a boiling point different from the boiling point of the original mixture which will allow separation to occur.

BALLING HYDROMETER OR BRIX HYDROMETER: a triple-scale wine hydrometer designed to record the specific gravity of a solution containing sugar.

BATCH FERMENTATION: fermentation conducted from start to finish in a single vessel.

BATF: Bureau of Alcohol, Tobacco, and Firearms; under the U.S. Department of Treasury. Responsible for the issuance of permits, both experimental and commercial, for the production of alcohol.

175

BEER: the product of fermentation by microorganisms; the fermented mash, which contains about 11-12% alcohol; usually refers to the alcohol solution remaining after yeast fermentation of sugars.

BEER STILL: the stripping section of a distillation column for concentrating ethanol.

BEER WELL: the surge tank used for storing beer prior to distillation.

BIOMASS: plant material, includes cellulose carbohydrates, ligniferous constituents, etc.

BOILING POINT: the temperature at which the transition from the liquid to the gaseous phase occurs in a pure substance at fixed pressure.

BRITISH THERMAL UNIT (BTU): the amount of heat required to raise the temperature of one pound of water one degree Fahrenheit under stated conditions of pressure and temperature (equal to 252 calories, 778 foot-pounds, 1,055 joules, and 0.293 watt-hours); it is the standard unit for measuring quantity of heat energy.

CALORIE: the amount of heat required to raise one gram of water one degree centigrade.

CARBOHYDRATE: a chemical term describing compounds made up of carbon, hydrogen, and oxygen; includes all starches and sugars.

CARBON DIOXIDE: a gas produced as a by-product of fermentation; chemical formula is CO_2.

CELLULASE: an enzyme capable of splitting cellulose.

CELSIUS (CENTIGRADE): a temperature scale commonly used in the sciences; at sea level, water freezes at 0° C and boils at 100° C.

CENTRIFUGE: a rotating device for separating liquids of different specific gravities or for separating suspended colloidal particles according to particle-size fractions by centrifical force.

COLUMN: vertical, cylindrical vessel used to increase the degree of separation of liquid mixtures by distillation or extraction.

CONCENTRATION: ratio of mass or volume of solute present in a solution to the amount of solvent.

CONDENSER: a heat-transfer device that reduces a thermodynamic fluid from its vapor phase to its liquid phase.

CONTINUOUS FERMENTATION: a steady-state fermentation system that operates without interruption; each stage of fermentation occurs in a separate section of the fermenter, and flow rates are set to correspond with required residence times.

COOKER: a tank or vessel designed to cook a liquid or extract or digest solids in suspension; the cooker usually contains a source of heat; and is fitted with an agitator.

DEHYDRATION: the removal of 95% or more of the water from any substance by exposure to high temperature.

DENATURANT: a substance that makes ethanol unfit for human consumption.

DENATURE: the process of adding a substance to ethyl alcohol to make it unfit for human consumption; the denaturing agent may be gasoline or other substances specified by the Bureau of Alcohol, Tobacco, and Firearms.

DEPARTMENT OF ENERGY: in October 1977, the Department of Energy (DOE) was created to consolidate the multitude of energy-oriented government programs and agencies; the Department carries out its mission through a unified organization that coordinates and manages energy conservation, supply development, information collection and analysis, regulation, research, development and demonstration.

DESICCANT: a substance having an affinity for water; used for drying purposes.

DEXTROSE: the same as glucose.

DISTILLATE: that portion of a liquid which is removed as a vapor and condensed during a distillation process.

DISTILLATION: the process of separating the components of a mixture by differences in boiling point; a vapor is formed from the liquid by heating the liquid in a vessel and successively collecting and condensing the vapors into liquids.

DISTILLERS DRIED GRAINS (DDG): the dried distillers grains by-product of the grain fermentation process which may be used as a high-protein (28%) animal feed. (See Distillers Grain.)

DISTILLERS DRIED GRAINS WITH SOLUBLES (DDGS): a grain mixture obtained by mixing distillers dried grains and distillers dried solubles.

DISTILLERS DRIED SOLUBLES (DDS): a mixture of water-soluble oils and hydrocarbons obtained by condensing the thin stillage fraction of the solids obtained from fermentation and distillation processes.

DISTILLERS FEEDS: primary fermentation products resulting from the fermentation of cereal grains by the yeast *Saccharomyces cerevisiae.*

DISTILLERS GRAIN: the nonfermentable portion of a grain mash comprised of protein, unconverted carbohydrates and sugars, and inert material.

ENRICHMENT: the increase of the more volatile component in the condensate of each successive stage above the feed plate.

ENZYMES: the group of catalytic proteins that are produced by living microorganisms; enzymes mediate and promote the chemical processes of life without themselves being altered or destroyed.

ETHANOL: C_2H_5OH: the alcohol product of fermentation that is used in alcoholic beverages and for industrial purposes; chemical formula blended with gasoline to make gasohol; also known as ethyl alcohol or grain alcohol.

ETHYL ALCOHOL: also known as ethanol or grain alcohol; see Ethanol.

EVAPORATION: conversion of a liquid to the vapor state by the addition of latent heat of vaporization.

FAHRENHEIT SCALE: a temperature scale in which the boiling point of water is 212° and its freezing point 32°; to convert °F to °C, subtract 32, multiply by 5, and divide the product by 9 (at sea level).

FEEDSTOCK: the base raw material that is the source of sugar for fermentation.

FERMENTABLE SUGAR: sugar (usually glucose) derived from starch and cellulose that can be converted to ethanol (also known as reducing sugar or monosaccharide).

FERMENTATION: a microorganically mediated enzymatic transformation of organic substances, especially carbohydrates, generally accompanied by the evolution of a gas.

FERMENTATION ETHANOL: ethyl alcohol produced from the enzymatic transformation of organic substances.

FRACTIONAL DISTILLATION: a process of separating alcohol and water (or other mixtures).

FUSEL OIL: a clear, colorless, poisonous liquid mixture of alcohols obtained as a by-product of grain fermentation; generally amyl, isoamyl, propyl, isopropyl, butyl, isobutyl alcohols and acetic and lactic acids.

GASOHOL (Gasahol): registered trade names for a blend of 90% unleaded gasoline with 10% fermentation ethanol.

GASOLINE: a volatile, flammable liquid obtained from petroleum that has a boiling range of approximately 29°-216° C and is used as fuel for spark-ignition internal combustion engines.

GELATINIZATION: the rupture of starch granules by temperature which forms a gel of soluble starch and dextrins.

GLUCOSIDASE: an enzyme that hydrolyzes any polymer of glucose monomers (glucoside). Specific glucosidases must be used to hydrolyze specific glucosides; e.g., B-glucosidases are used to hydrolyze cellulose; *a*-glucosidases are used to hydrolyze starch.

HEAT EXCHANGER: a device that transfers heat from one fluid (liquid or gas) to another, or to the environment.

HYDRATED: chemically combined with water.

HYDROCARBON: a chemical compound containing hydrogen, oxygen, and carbon.

HYDROLYSIS: the decomposition or alteration of a polymeric substance by chemically adding a water molecule to the monomeric unit at the point of bonding.

HYDROMETER: a long-stemmed glass tube with a weighted bottom; if floats at different levels depending on the relative weight (specific gravity) of the liquid; the specific gravity of other information is read where the calibrated stem emerges from the liquid.

INOCULUM: a small amount of bacteria produced from a pure culture which is used to start a new culture.

LEADED GASOLINE: gasoline containing tetraethyllead to raise octane value.

LIQUEFACTION: the change in the phase of a substance to the liquid state; in the case of fermentation, the conversion of water-insoluble carbohydrate to water-soluble carbohydrate.

MALT: barley softened by steeping in water, allowed to germinate, and used especially in brewing and distilling.

MASH: a mixture of grain and other ingredients with water to prepare wort for brewing operations.

MEAL: a granular substance produced by grinding.

METHANOL: a light volatile, flammable, poisonous, liquid alcohol, CH_3OH, formed in the destructive distillation of wood or made synthetically and used especially as a fuel, a solvent, an antifreeze, or a denaturant for ethyl alcohol, and in the synthesis of other chemicals; methanol can be used as fuel for motor vehicles; also known as methyl alcohol or wood alcohol.

METHYL ALCOHOL: also known as methanol or wood alcohol; see Methanol.

OCTANE NUMBER: a rating which indicates the tendency to knock when a fuel is used in a standard internal combustion engine under standard conditions.

PACKED DISTILLATION COLUMN: a column or tube constructed with a packing of ceramics, steel, copper, or fiberglass-type material.

pH: a term used to describe the free hydrogen ion concentration of a system; a solution of pH 0 to 7 is acid; pH of 7 is neutral; pH over 7 to 14 is alkaline.

PLATE DISTILLATION COLUMN (Sieve tray column): a distillation column constructed with perforated plate or screens.

PROOF: a measure of ethanol content; 1 percent equals 2 proof.

PROOF GALLON: a U.S. gallon of liquid which is 50% ethyl alcohol by volume; also one tax gallon.

PROTEIN: any of a class of high molecular weight polymer compounds comprised of a variety of amino acids joined by a peptide linkage.

RECTIFICATION: with regard to distillation, the selective increase of the concentration of the lower volatile component in a mixture by successive evaporation and condensation.

RECTIFYING COLUMN: the portion of a distillation column above the feed tray in which rising vapor is enriched by interaction with a countercurrent falling stream of condensed vapor.

REFLUX: that part of the product stream that may be returned to the process to assist in giving increased conversion or recovery.

RENEWABLE RESOURCES: renewable energy; resources that can be replaced after use through natural means; example: solar energy, wind energy, energy from growing plants.

SACCHARIFY: to hydrolyze a complex carbohydrate into a simpler soluble fermentable sugar, such as glucose or maltose.

SACCHAROMYCES: a class of single-cell yeasts which selectively consume simple sugars.

SIGHT GAUGE: a clear calibrated cylinder through which liquid level can be observed and measured.

SOLAR ENERGY RESEARCH INSTITUTE (SERI): the Solar Energy Research Development and Demonstration Act of 1974 called for the establishment of SERI, whose general mission is to support DOE's solar energy program and foster the widespread use of all aspects of solar technology, including direct solar conversion (photovoltaics), solar heating and cooling, solar thermal power generation, wind conversion, ocean thermal conversion, and biomass conversion.

SPECIFIC GRAVITY: the ratio of the mass of a solid or liquid to the mass of an equal volume of distilled water at 4° C.

SPENT GRAINS: the nonfermentable solids remaining after fermentation of a grain mash.

STARCH: a carbohydrate polymer comprised of glucose monomers linked together by a glycosidic bond and organized in repeating units; starch is found in most plants and is a principal energy storage product of photosynthesis; starch hydrolyzes to several forms of dextrin and glucose.

STILL: an apparatus for distilling liquids, particularly alcohols; it consists of a vessel in which in liquid is vaporized by heat, and a cooling device in which the vapor is condensed.

STILLAGE: the nonfermentable residue from the fermentation of a mash to produce alcohol.

STOVER: the dried stalks and leaves of a crop remaining after the grain has been harvested.

STRIPPING SECTION: the section of a distillation column below the feed in which the condensate is progressively decreased in the fraction of more volatile component by stripping.

VACUUM DISTILLATION: the separation of two or more liquids under reduced vapor pressure; reduces the boiling points of the liquids being separated.

VAPORIZE: to change from a liquid or a solid to a vapor, as in heating water to steam.

VAPOR PRESSURE: the pressure at any given temperature of a vapor in equilibrium with its liquid or solid form.

WORT: the liquid remaining from a brewing mash preparation following the filtration of fermentable beer.

YEAST: single-cell microorganisms (fungi) that produce alcohol and CO under anaerobic conditions and acetic acid and CO under aerobic conditions; the microorganism that is capable of changing sugar to alcohol by fermentation.

Acknowledgments

Writing this book has been an enjoyable experience, mostly because of the many fine people who have helped keep me on the right track and the many others who, by their encouragement, have pushed me onward.

There may be a few places where my recommendations differ slightly from those provided by experts more knowledgeable than myself. That means that, by acknowledging their assistance, I don't intend to imply that they endorse this book. I am simply grateful.

Among those I want to thank for their valuable technical assistance are: Dr. T.P. Lyons of BIOCON, U.S; Charles Jocelyn of Arnold Edwards Corporation; Jack Dubrul, Dale Fillion and Wolfgang White of The Automaster; Archie Zeithamer of Alexandria, Minnesota; Vincent DeSalvio, Howard D. Criswell Jr. and Robert E. Daugherty of the Bureau of Alcohol, Tobacco and Firearms; John Lincoln of Stonington, Connecticut; Gary Husted of the University of Vermont; and Albert Turner of Selma, Alabama.

Additional helpful information has come from one or more individuals from many corporations and associations, notably these: Novo Laboratories of Wilton, Connecticut; Miles Laboratories of Elkhart, Indiana; Anheuser-Busch of St. Louis, Missouri; Archer-Daniels Midland of Decatur, Illinois; Iowa Corn Promotion Board of Des Moines; National Gasohol Commission of Lincoln, Nebraska; National Alcohol Fuels Information Center of Golden, Colorado; Joseph E. Seagrams of Lexington, Kentucky; and South West State University of Marshall, Minnesota.

I am especially grateful to the Solar Energy Research Institute of Golden, Colorado for permission to reproduce sections of the appendices appearing in the Institute's excellent book, *Fuel From Farms — A Guide to Small Scale Ethanol Production*. Similarly, I'm indebted to Joseph Earl Dabney for his delightful chapter, "The ABCs of Pure Corn," and to Oren Long for his overview of the Zeithamer still.

Almost all of the photographs appearing in Chapters 2 and 3 were taken by Glenn Moody, a good friend and fine professional photographer from Stowe, Vermont. Historical photos were provided by Joseph E. Dabney, the Library of Congress Wittemann Collection, the Internal Revenue Service and the Bureau of Alcohol, Tobacco and Firearms. Photos of the Zeithamer still are courtesy of the Iowa Corn Promotion Board.

To those who have encouraged me along the way, I give hearty thanks: Keith Bondesen, who kept the mash stirring; Chuck Armel and Roger Cota, who thoughtfully critiqued still designs; Andy Lombard, who kept friendly track of progress; Peggy McGowan, who provided editorial assistance; Bob Bennett and Bob Neff, who kept the clippings coming; Jack Williamson, who backed the project all the way; David Robinson, who improved my punctuation and spelling; and, most important, Roger Griffith, who asked about 16 good questions a day.

Finally, I am deeply indebted to my daughter, Lizzie, and my wife, Mary Esther. Although Lizzie's mother used an expletive here and there, she had the good sense not to divorce me while I spent several months puttering away in the garage, working with my still.

Index